Discoveries
and Inventions
That Changed Our World

Discoveries and Inventions
That Changed Our World

The pioneers of modern science—
what they did and why it matters

Pete Moore

A QUINTET BOOK

Published by Apple Press
Sheridan House
112-116A Western Road
Hove
East Sussex BN3 1DD

Printed 2005

ISBN 1-84543-097-2

This book was designed and produced by
Quintet Publishing Limited
6 Blundell Street
London N7 9BH

EMC2

Project Editor: Corinne Masciocchi
Editor: Nicole Foster
Designer: Roger Fawcett-Tang

Creative Director: Richard Dewing
Publisher: Oliver Salzmann

Color separation by Universal Graphics Pte Ltd, Singapore
Printed in China by Midas Printing International Limited

Medical disclaimer
All statements and information given in this book regarding
theories and techniques should not be construed as medical
instruction.

Acknowledgments
Pete Moore wishes to thank Jim Al-khalili, Tim Berners-Lee, Carl
Djerassi, Ted Nelson, and Ilya Prigogine for their help in checking
the factual content of some of the entries in this book.

Contents

When Einstein announced that $E=mc^2$ the world did not change. Nothing had been invented. The way that particles behave was unaltered. For that matter, the universe stayed just as it always had been. But we did change. As the intricacies of quantum physics revealed themselves to inquiring minds, humanity's view of the entire universe was altered forever. Atoms, electrons, and sub-atomic particles were no different from what they had been for billions of years – the only real difference was that we knew about them.

This book takes a look at over 70 of the key names in the history of thought: the people who have made us change the way we look at our universe and ourselves. It uncovers their backgrounds, their personal motives, and the contributions that they have made – their intellectual legacy. The intention is to reveal each person's chief achievements, while acknowledging that these are supported by many other discoveries and data that occupy packed shelves of academic libraries.

On the whole, these thinkers had two key aims. Those like astronomer Edwin Hubble and his friends quite simply wanted to discover what exists and how everything fits together. On the other hand, scientific inventors like James Watt and Thomas Edison set out to manipulate the fundamental processes involved in the universe in order to create tools and techniques that could make life that little bit more enjoyable.

The people illustrated in this book are a motley crowd of extremists. Many of them lived in relative obscurity, and only entered the glamorous limelight of fame after they died. As children, most failed to fit into the straitjacket of standard education. They either "excelled" and, like physicist Abdus Salam, had qualifications way ahead of their peers, or were classified as failures. Isaac Newton was described by his teacher as idle and inattentive. Thomas Edison played truant. Psychiatrist John Watson was

Introduction

sometimes violent. In retrospect we can see that they were probably just too clever to be interested in standard schooling and too creative to accept information without questioning it. Still others, like the explorer of gases Robert Boyle, were ill as children and had little formal education. They are examples that should encourage any parent who sees their child struggling at school.

It's interesting to note how many of the great achievements occurred against a backdrop of war, conflict, and political unrest. The ancient Greek philosophers like Socrates were partly driven by a need to make sense of a feuding world, and the twentieth-century sages of particle physics like Heisenberg and Niels Bohr found their science being used in warfare. Because of their intelligence or specific expertise, others like Plato and Henry Ford found themselves unwittingly on the frontlines of policy. Still others, such as Erwin Schrödinger, experienced life literally in the line of fire.

Each of the characters in this collection, however, is known because at some point in their lives they were in the right place at the right time. Most people know the name of Charles Darwin, but fewer recognize Alfred Wallace. While both more or less simultaneously came to the same conclusion about evolution, Darwin had wealth and political friends and, living in England, won the race to publish his thoughts. Some, like the so-called father of the contraceptive pill Carl Djerassi, gained fame because their discovery fitted in with a tide of socio-political change. Countless other scientists and explorers of the mind or body have probably had great ideas but got nowhere with them because they were too far ahead of their time.

The entries in this book also show the extent to which these explorers of reality have struggled not only with their science, but also with questions about how their findings fit with religious beliefs. Some viewed this in terms of conflict, others compatibility, but the development of science has necessarily caused individuals and societies to reconsider basic patterns of behavior and has altered the way everyone lives.

The mysterious equation $E=mc^2$ formed a landmark for science, but visual proof of Einstein's prediction that light would bend as it passed next to the Sun captured the imagination of the public. The equation was intangible; the photographic evidence of stars apparently moving in space was easier to grasp. This book shines light on the complexities of thought that underpin some of the more accessible aspects of each great thinker. As such it forms the essential lay reader's guide to the greatest and most influential thinkers of human history.

Pete Moore

The world around us

THE STRUCTURE OF NATURE

Democritus
Robert Boyle
Antoine Lavoisier
John Dalton
Edward Frankland
Dmitri Mendeleev
Marie Curie
Ernest Rutherford

Niels Bohr
Linus Pauling
Dorothy Crowfoot
Hodgkin
Richard Feynman

THE EARTH

Charles Lyell
Alfred Wegener
James Lovelock

THE UNIVERSE

Edwin Hubble
George Gamow
Fred Hoyle
Ilya Prigogine

MATHEMATICS

George Boole
David Hilbert
Norbert Weiner
Alan Turing
Claude Shannon
René Thom
Benoit Mandelbrot
John Nash

INVENTION AND INNOVATION

Leonardo da Vinci
James Watt
Thomas Edison
Henry Ford
Ted Nelson
Tim Berners-Lee

Thales

c.624–c.545 B.C.

Acquaintances
— Anaximander (611–545 B.C.)
— Pythagoras (c.580–500 B.C.)

History records people who make changes either to the way things are, or the way we perceive them. Thales is remembered for the latter, because he disregarded the classical mythical understanding of life, the universe, and everything, in favor of a more physical understanding. While later work shows errors in his concepts, the basic principle that you can understand things by studying what they are made of stands and forms the foundation for future philosophy and science. Like other thinkers of ancient Greece he wasn't confined to an ivory tower, but engaged in disciplines as varied as engineering and statesmanship.

Water is not just wet

The search for origins occupies many explorers. For geographers it may be the origins of a river, for philosophers it is the origins of existence. As Thales looked out on his world he was dissatisfied with the mythical explanations that provided good stories, but made no attempt to grapple with the physical nature of the earth and all that lived on it. He rejected the widely held idea that the gods were part of everything. Instead, he went in search of physical explanations.

Born in Miletos, a town on the Aegean coast of what is now Turkey, Thales had a belief that there must be a fundamental building block, a substance from which everything else was made. He concluded that this was water. Water after all was essential for life. Drink it and you grow, take it away and everything dies. Water, he thought, could readily be made very fine, in which case it becomes air, or alternatively it could be compacted to a slime that becomes earth.

Not only was everything made of this elemental substance, but all life was supported on it—literally. The Earth, said Thales, floated on water. There were two compelling strands of evidence. First, it couldn't be supported by air, because air was incapable of supporting anything, but water could hold up large objects like ships and logs. Secondly, you could observe the effects of the Earth floating on water in that on occasions it rocked suddenly—an earthquake. Quite obviously this was caused by the water's movement.

Of pyramids and shadows

While Thales spent most of his life in Greece, there are plenty of reports of him traveling in the hope of extending his scholarship. Legend has it that while in Egypt he developed a system for measuring the height of the pyramids. It dawned on him that at certain times of day his shadow was the same length as his height. He concluded that at the same moment, the height of the pyramid would also be the same as the length of its shadow.

As an observation it was interesting in itself, but all the more important because he was aware that this process worked because the ratio of the length of the shadow and the height of the object was the same for all objects if measured at the same time of day. In so doing he became one of the founding fathers of geometry, even though there is no indication that he followed up the exercise with any mathematical calculations.

Staring at stars

Not content with measuring things that are bound to the earth, Thales was also intrigued by the sky and started to realize that the sun, moon, and stars moved with remarkable regularity. He is attributed with being the first person to spot the grouping of stars now known as Ursa Minor, or the Bear, and to use it in navigation. More dramatically he also predicted the May 28, 585 B.C., eclipse of the Sun. This again gave weight to his theory that the universe was operated according to physical rules, rather than at the whim of a myriad of gods.

According to one probably apocryphal story, Thales was the first-recorded absent-minded academic. Once, he was so engrossed in his observation of the stars, he walked backward and fell into either a ditch or a well. Rescue was at hand in the form of Theodorus, a witty and attractive servant girl who inquired, "How do you expect to understand what is going on up in the sky if you do not even see what is at your feet?"

Nothing remains of Thales' original work and all our knowledge of him is derived from later philosophers. He was a keen athlete, and it is therefore fitting that, according to the third-century A.D. biographer Diogenes Laertios, he was born in the first year of the 35th Olympiad (640 B.C.) and died of heat exhaustion while watching the 58th Olympiad (548–545 B.C.).

Mottos of the seven sages of Greece

Pittacus of Mitylene (c.650–c.570 B.C.)
"Know thine opportunity"

Solon of Athens (638–559 B.C.)
"Know thyself"

Thales of Miletos (c.624–c.545 B.C.)
"Who hateth suretyship is sure"

Periander of Corinth (died 585 B.C.)
"Nothing is impossible to industry"

Chilon of Sparta (sixth century B.C.)
"Consider the end"

Bias of Prinene (sixth century B.C.)
"Most men are bad"

Cleobulus of Lindoe (sixth century B.C.)
"The golden mean" or *"Avoid extremes"*

Pythagoras

c.580–500 B.C.

Acquaintances

— Thales (c.624–c.545 B.C.)
— Anaximander (611–545 B.C.)
— Pherekydes (sixth century B.C.)

Understanding the world through mathematics was at the heart of Pythagoras' mission in life. Guided by teachers such as Thales, his goal was not so much to be clever at mathematics, but to find ways of making sense of everything he saw. He built on previous theories and found ways of proving long-held concepts, including producing an elegant proof of what is now known as Pythagoras' Theorem. Everything that he did was aimed at developing a way of life marked by moral asceticism and purification and, by establishing communities of students and co-believers, Pythagoras hoped to integrate his lifestyle with his learning.

First pure mathematician

In ancient Greece philosophers had a tendency to move from place to place, some because their families moved, others to increase their learning, and still others in order to escape persecution. At different times in his life, each of these reasons caused Pythagoras to pack his bags and board a boat. The result was that he fed on a rich diet of intellectual adventure.

As a late teenager he went to study with Thales, though by this time Thales was an old man and had handed over most of his lecturing to his former student Anaximander. From them Pythagoras gained an excitement about Egyptian concepts of geometry, and it was undoubtedly here that he first encountered a 1,000-year-old Babylonian theory about any triangle that had a right angle at one of its corners—the theory that school pupils now chant as "the square on the hypotenuse is the sum of the squares on the other two sides."

The theory is attributed to Pythagoras, not because he was the first to spot the relationship, but because he developed a system of proof: a method of showing mathematically why this is the case. It was probably no surprise to Pythagoras to find that the simple step-by-step application of basic principles could prove that these triangles shared fundamental properties, because he was convinced that there were mystic powers locked up in numbers. It was as if he thought they were in some sense alive. Numbers were either masculine or feminine, beautiful or ugly. He thought that 10 was the very best number and was pleased that it was made up of the first four integers, i.e. 1+2+3+4 = 10.

His searching caused him to discover that there were some numbers that could not be generated by multiplying two others together, for example, 1, 2, 3, 5, 7, 11... These became known as prime numbers, with the so-called composite numbers being the ones that could be generated by multiplying two others. For example 4 is the product of 2 x 2 or 6 of 2 x 3.

He was, however, puzzled by other numbers that didn't appear to exist. When he applied his theorem to a triangle that has two sides coming from the right angle that are both one unit long, the theorem breaks down. The square of 1 equals 1, so add two 1s together and you get 2. But Pythagoras had shown that 2 was a prime number. There was no way that he could find a number which when multiplied by itself came to 2 (a square root). This he said was because it was an irrational number, it did not fall into the rules of ratios that he had found.

Making music

Ratios, Pythagoras found, were not confined to the math class, they could also make music. As a child he had learnt to play the lute, and it was probably while practicing one day that he started to see what happened when he twanged strings of different lengths. If there was a beauty in mathematics, then applying mathematical principles to music should be pleasing to the ear.

After a bit of experimentation he noted that vibrating strings produce harmonious tones when the ratios of the lengths of the strings are whole numbers. This concept worked whatever instrument he tried it on. It seemed to be another universal principle.

above *A manuscript page in Latin translation dated eighth to ninth century A.D. of the work of the early Greek philosopher and mathematician Pythagoras.*

Spheres of influence

Thales along with all other thinkers was quite content to believe that the Earth was flat. This was quite obviously the case, because otherwise you would fall off. Pythagoras had a different idea, and was the first person to teach that the Earth was a sphere. It was a revolutionary idea, but one that made sense of the observations he made about the fact that the Sun, Moon, and Venus move in the sky.

As is so often the case with scientists, Pythagoras got the spherical bit right, but didn't grasp the whole story, placing the Earth at the center of the universe with everything rotating around it. He did, however, recognize that the orbit of the Moon was inclined to the equator of the Earth and he was one of the first people to realize that Venus seen as an evening star was one and the same planet as Venus the morning star.

Travel and teaching

As a youngster, Pythagoras had traveled with his father Mnesarchus, a merchant from the eastern Mediterranean town of Tyre. Having brought corn to the Greek city of Samos during a famine, the city fathers granted Mnesarchus citizenship and this, coupled with the fact that Pythagoras' mother, Pythais, had come from Samos, gave the opening for Pythagoras to spend time with the likes of Thales. It appears he followed his father's example and while living in Samos fell in love and married Theano, with whom he lived for the rest of his life.

As his fame and experience grew, Pythagoras was keen to find ways of integrating his learning into his life. He had seen the way that religious communities in Egypt sought purity in their thinking and lifestyles, and he copied some of their principles of secrecy and commitment when he set up his own school in Samos, which became known as "thē semicircle."

He and the Samians, however, failed to see eye to eye as they disliked the way that he tried to reduce everything to symbols, so around 518 B.C. he moved on to Crotone, a city on the east of the heel of southern Italy. There he established a community that did take root. His followers, the Pythagoreans, lived by a strict code of secrecy and were forbidden from talking to anyone outside the sect. Sadly, this means that much of their work was inevitably lost to future generations. They tried to live in harmony with all other animals, they ate no meat, and refused to wear anything made of animal skins. The inner circle of followers was known as mathematikoi and, unusually for the time, included women among the membership. An outer circle, the akousmatics, lived in their own homes and came in by day.

An underlying belief of the Pythagoreans was that philosophy can be used to enhance spiritual perception, and that the soul can rise to be in union with the divine. They also believed in reincarnation, a belief that may have fuelled the legend that Pythagoras committed suicide after having to flee from political instability in Italy. Whatever the cause of his death, his days ended in 500 B.C. in Metapontum, a town on what is now the south-east coast of Italy.

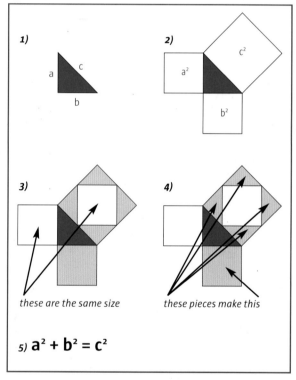

1)

2)

a c
b

a^2 c^2
b^2

3)

these are the same size

4)

these pieces make this

5) $a^2 + b^2 = c^2$

above *Pythagoras proved that a square drawn so that its sides are the length of the longest side of a right angle triangle has the same area as the two squares drawn on the two opposing sides.*

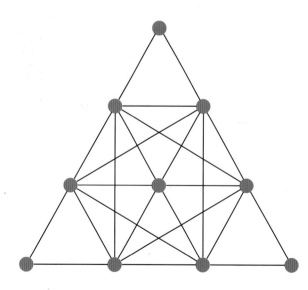

left *Ten was seen as a number of beauty and Pythagoras enjoyed the concept that the sum of the first four numbers equaled ten as he demonstrated in this pyramid of dots.*

PLATO

RAPH. SANCT. Vrb. pinxit in aed. Vatic.

c.428–c.348 B.C.

Acquaintances
—Socrates (469–399 B.C.)
—Aristotle (384–322 B.C.)
—Dionysius II (c.397–344 B.C.) and his brother Dion

"The safest general characterization of the European philosophical tradition is that it consists of a series of footnotes to Plato." So said the English mathematician and philosopher Alfred North Whitehead (1861–1947). Central to Plato's thinking is that people should seek virtue by studying what he called the Good, a non-physical absolute concept that never changes. If you know Good, you will live well because your thoughts and desires will automatically be shaped by that knowledge. As a student of Socrates and teacher of Aristotle, Plato stands at a pivotal point in philosophical history.

Philosophy rests on broad shoulders

Among thinkers, Plato was a physical as well as an intellectual giant. Popular tradition has it that he was originally named Ariston, but picked up the nickname of Plato, meaning "broad," at school. This apparently referred to the size and power of his shoulders, which he put to good use when he won the Isthmian wrestling competition. He is, however, remembered for his mental agility and his ability to grapple with ideas.

As a young man, Plato had political ambitions. He had been born into an aristocratic family. His father, Ariston, had descended from the early kings of Athens and his mother, Perictone, from the sixth-century B.C. legislator Solon. Quite soon he became disillusioned with the poor state of Athenian politics that resulted from the mess that had been the Peloponnesian War.

Being introduced to Socrates radically altered Plato's life. Socrates taught a method of thinking that elucidated truth through a series of questions and answers. It became a central feature of much of Plato's prolific writing. The Socratic method was to ask for definitions of familiar concepts, like "justice" and "courage," and then probe the definition by asking a series of questions. His intention was to lead people to start contradicting themselves and in so doing uncover any weakness in the initial definition. Socrates was convinced that the wisest people were those who were aware of how little they knew. He also pointed Plato on his quest to define the nature of virtue. To have virtues is to be a good specimen of human kind.

Reanimating Socrates

In 399 B.C. Socrates had fallen foul of a newly established democratic government. Put on trial for corrupting youth he was sentenced to death and elected to drink hemlock rather than arguing in favor of a fine or accepting the offer of help to escape. His rationale for this obstinacy was that he believed that doing harm damaged one's soul. Any action that seemed underhand or dishonest would be detrimental to his soul. As the soul survives death it would be better to die.

Not surprisingly, Plato decided that Athens was a dangerous place to live, and he spent the next 12 years wandering around the Mediterranean, visiting Italy, Sicily, and Egypt — meeting philosophers, priests, and prophets, and writing plays. In his plays he used Socrates as a character, bringing his mentor back from the grave, and throwing light on his concepts. In *Gorgias*, Plato portrays Socrates confronting Polus, a young man who holds that

immoral acts can bring the greatest amount of pleasure, measuring actions in terms of their immediate material outcome. Socrates disagrees. Whatever the immediate pleasure, he says, immorality will damage the soul.

Of Forms, Good, and a three-part personality

At the age of 40, Plato returned to Athens and set up the Academy, arguably Europe's first university. The curriculum covered a broad range of subjects, including astronomy, biology, mathematics, political theory, and philosophy, and ran until Emperor Justinian closed it in 529 A.D. Plato's vision was to train a new generation of "philosopher-kings" who would lead the Greek world.

While he was running the Academy he was developing his theory of Forms or "ideas." Forms, according to Plato, are timeless, unchanging, immutable, and universal; they are not created nor do they create anything. They simply exist. This contrasts with "Particulars," that are objects of perception, opinions, or beliefs. These, on the other hand, may change with time.

In the parable of the cave, told in arguably his best-known work, *The Republic*, Plato challenges people's notion of reality. He sets up a situation in which he likens the general population to people who have been chained immobile in a cave since childhood. Behind them is a fire, and they are shackled so that they can only look in front of themselves. Consequently, their only view of life is of shadows cast by the light of the fire, that dance on the cave wall in front of them. As far as they can see, these shadows are "reality." But then one of these people, the philosopher, breaks free. He stands up, looks around and sees the real people and the fire that casts the shadows. In this shocking discovery, he realizes that there is more to life than shadows. It's not a pleasant experience, because looking directly at the fire hurts his eyes. He then walks out of the cave and sees the sun. Seeing this source of light is more painful still, but for the first time he sees the perfect source of light. In this analogy, the sun is the Form of Good, the highest of the Forms.

The philosopher is faced with two problems. First he has the pain of his own investigation, and secondly, when he goes back into the cave to relate his findings to the other captives he is likely to meet with disbelief, rejection, and even anger.

Plato argued that everything we see and call beautiful, in some way resembles the Form of Beauty. Two people independently come to the conclusion that a person or object is beautiful because they both recognize the Form of

above *Plato disputing with his students in the garden of the Academy.*

Beauty. In a similar way, everything that we see as "just" resembles the Form of Justice. Disputes about the rightness of actions then depend on how well the outcome will conform with the Form of Good.

For a person to act justly requires that while they seek the Form of Good, they keep the three parts of their personality in balance. The person needs wisdom, which comes from reason; courage, which comes from the spirited part of man; and self-control, which rules the passions.

The Dionysius disaster

Plato believed that if you educate a person so that he or she can see that a particular action is not good for them, then they will not perform that action. This knowledge will enable them to make good decisions and to rule wisely.

In *The Republic* he expands the idea, building a story of a Utopian society ruled by a philosopher-king who has mastered the discipline of "dialectic" and studied the hierarchy of Forms. The society is organized into a rigid hierarchy in which workers, soldiers, and rulers all know their relative positions and there is communism of property and family. The rulers have totalitarian powers.

In 367 B.C. he had the opportunity to put these ideas to the test when he was invited to visit Syracuse and teach its ruler Dionysius II how to be a philosopher-king. The experiment failed. Plato paid two visits, but the second trip was such a disaster that he had to run for his life. He died in his sleep 20 years later.

Timeline

c.428 B.C.
Born in Athens to an aristocratic family

c.409 B.C.
Plato becomes a student of Socrates, meeting him through his mother's uncle, Critias, and his mother's brother, Charmides, who were Socrates' friends

c.399 B.C.
After the Greeks executed Socrates, Plato goes to live in Megara where he took refuge with the philosopher Euclides, before starting to travel around Greece, Egypt, and the Greek cities in southern Italy
During these travels he encounters the Pythagoreans and makes friends with Dion, brother-in-law of Dionysius I, the ruler of Syracuse. As he moved from place to place, he also wrote his first group of "dialogues," which include *Apology*, the *Crito*, *Charmides*, *Euthyphro*, *Laches*, *Lysis*, *Hippas Minor*, *Hippas Major*, *Gorgias*, *Ion*, and *Protagoras*

c.387 B.C.
Returns to Athens and founds the Academy, which became a famous center of philosophical, mathematical, and scientific research

c.367 B.C. and c.361–360 B.C.
Visits Sicily again and attempts to train Dionysius II to be a philosopher-statesman, but fails

c.348 B.C.
Plato dies in his sleep

Even among revolutionary thinkers, a person with new ideas is not always accepted. While studying with Plato, Aristotle disagreed with some of his ideas and so was effectively banished from Plato's Academy. This setback liberated him to pursue his ideas unrestricted by Socratic theories that non-physical Forms such as Truth and Beauty were the keys to understanding. Instead, he developed his theories by employing experiment and observation. Aristotle maintained that sense, experience, and reason were all necessary for knowledge. His importance lies as much in his analytical method as in the conclusions he reached, though his work still underpins both Christian and Islamic philosophy.

Aristotle

384–322 B.C.

Acquaintances
- Plato (c.428–c.348 B.C.)
- Speusippus (c.407–339 B.C.)
- King Amyntas of Macedonia (reigned c.393–370 B.C.)
- Proxenus (4th century B.C.)
- Philip of Macedonia (c.382–336 B.C.)
- Alexander the Great (c.356–323 B.C.)

Of form, substance, and society

"Plato is dear to me, but dearer still is truth," said Aristotle. It was an attitude that would cause him pleasure and pain. Pleasure because he enjoyed working with Plato and studying his methods, but pain, because he came to believe that, in some crucial ways, Plato was wrong. Aristotle found that Plato's theory of Forms was insufficiently grounded in reality. Instead of believing that everything is ultimately related to an absolute truth, he developed methods of analysis and categorization that were based on observing the substance from which things were made.

Aristotle's starting point was that everything had a purpose. It might on occasions be difficult to see what the purpose is, and sometimes the purpose might be to deliberately do nothing—but there will be purpose.

To study a situation, or an action, he would categorize it into a series of subordinate and superior aims. The subordinate aims are needed to fulfill the superior. For example, striking a match is needed to light a fire. Lighting a fire is needed to warm a room. Warming a room is needed to dry washed clothing, and so on. The place where this converges with Plato's ideas is that for Aristotle, the pinnacle of this tower of superiority is the Good.

According to Aristotle, humans by nature are social and moral, therefore all aims eventually lead to the Good, not necessarily of the individual but of humanity as a whole. He then defines the supreme Good as "happiness." For Aristotle everyone is part of a group, be it a family, village, town, or city state. There is no place for individualism or free-thinkers. Our individuality is defined within the group to which we belong. Consequently, it is the happiness of the group that is of paramount importance, because without that the individual cannot be happy. This then is Aristotle's fundamental argument in favor of politicians studying ethics—so that they can enable the group to be happy, and therefore enable all members of society to have a good life.

As a consequence of his emphasis on the community as opposed to the individual, he believed that hierarchy and subordination were inevitable and as a result thought that slavery was a very normal part of a well-ordered society.

Dividing the soul, and virtue

Aristotle tried to gain understanding by dividing things into constituent parts, and he took a similar approach when he studied the soul. The soul, he said, was composed of two broad components: a rational part and an irrational part. The rational half was itself subdivided into "scientific" and "calculative" sections, and the irrational half was made of a "desiderative" and a "vegetative" part.

A person operates by combining the workings of the four sections. The vegetative part, for example, may require food. The desiderative part may want lots of chocolate rather than bread, but the scientific part knows that too much chocolate will be bad for teeth and weight. The calculative part works out if there is a possible solution and attempts to strike a compromise: chocolate spread on a slice of bread. The scientific part can then take over and make it happen. Problem solved, and life continues.

Like Plato, Aristotle believed that seeking "virtue" was important. In his book *Politics* Aristotle comments, "When devoid of virtue, man is the most unscrupulous and savage of animals, and the worst in regard to sexual indulgence and gluttony." Enjoying placing things in categories, he expanded Plato's concept by dividing virtues into two groups—the so-called 12 "moral" and nine "intellectual" virtues. He believed that each moral virtue lay between the non-virtuous extremes of excess and deficiency. For example, the virtue of courage lay between excessive rashness and deficient cowardice, and the virtue of modesty came between shyness and shamelessness. His intellectual virtues consisted of art, scientific knowledge, prudence, intelligence, wisdom, resourcefulness, understanding, judgment, and cleverness. Again, each of these lay between the extremes of excess and deficiency. His doctrine of the mean encouraged people to find the virtuous middle way.

Categorizing life

Aristotle's desire to place things in groups led him to the conclusion that there must be basic commodities that combine to make all things. His basic four groups were earth, air, fire, and water. His love of categories also led him to divide all people into three clusters. There was the majority group who love pleasure, a smaller group—which includes politicians—who love honor, and the smallest, but most elite, group who love contemplation. These were the philosophers.

above *The four elements theory (fire, water, air, and earth) was first suggested by the Greek philosopher Empedocles in the fifth century B.C. He claimed that all materials were made from various combinations of these elements. The theory was elaborated upon by Aristotle, who claimed motion was caused by each element's desire to reach its rightful location.*

The next task was to find the key feature that distinguished human beings from other animals. His answer was our ability to reason. Reasoning was important because it was the only way to help us sort out what is good, and without that there is no way that we will find happiness.

Having established that division he set about categorizing the entire biological world, or as much of it as he could get hold of. He grouped animals with similar characteristics into genera and then divided these genera into species. The same process is still applied today, though subsequent research has caused some of the individuals to be moved around. His main grouping, however, defined animals according to whether they had blood, which effectively divided them into the group now known as vertebrates, while the bloodless ones are the invertebrates.

Fascinated by the physical nature of living things Aristotle performed dissections so that he could try to make sense of how they worked. He described how a chick develops within an egg and worked out that dolphins and whales are different from fish. He noted that ruminant animals like cows had stomachs composed of a number of different chambers, a feature that distinguished them from other simple-stomached animals like pigs, dogs, and man. Not content with observing the make-up of large animals, he also studied the social organization of bees.

A legacy of influence

The mark of a philosopher is in the impact the thinker has on future generations, and by this assessment Aristotle passes the test with flying colors. His ideas are picked up by the 12th-century Islamic scholar Averroës, and consequently influence Islamic philosophy, and are later adopted by the Medieval philosopher Thomas Aquinas in the 13th century, whose work and concept of Natural Law underpin much thinking in the "Christian world."

Timeline

c.384 B.C.
Born in the Greek colony of Stagira, Macedonia, Aristotle was the son of Nicomachus, the court physician to the king of Macedonia

c.367 B.C.
Nicomachus died while Aristotle was young, but when Aristotle is about 18, he goes to Athens, initially as a pupil in Plato's Academy, but later becomes a teacher

c.347 B.C.
After Plato's death Speusippus succeeded Plato as head of the Academy. Aristotle leaves and goes to the kingdom of Atarneus in Asia Minor, where he marries the king's niece

c.342 B.C.
Aristotle is appointed tutor to Alexander, the son of Philip of Macedon, who went on to became Alexander the Great

c.335 B.C.
Aristotle returns to Athens and founds his own school, the Lyceum. This was built near to the temple of Apollo Lycieus

c.323 B.C.
Alexander the Great died.
Aristotle was alarmed at the anti-Macedonian sentiment that ran through Athens after Alexander's death. Accused of impiety and being fully aware of what had happened to Socrates he decides to make himself scarce. He takes refuge at Calcis in Euboea

c.322 B.C.
Aristotle dies in Euboea

Table of virtues

Deficiency	Virtue	Excess
cowardice	courage	rashness
insensibility	temperance	licentiousness
(indifference)	(restraint or moderation)	(disregarding convention, unrestrained)
illiberality	liberality	prodigality
(mean)	(giving and generous)	(recklessly wasteful, extravagant)
pettiness	magnificence	vulgarity
humble-mindedness	high-mindedness	vanity
lack of ambition	proper ambition	over-ambition
lack of spirit	patience	irascibility (easily angered)
understatement	truthfulness	boastfulness
boorishness	wittiness	buffoonery
cantankerousness	friendliness	obsequiousness
shamelessness	modesty	shyness
malicious enjoyment	righteous indignation	envy / spitefulness

above *According to Aristotle there are 12 moral virtues, each of which falls between two vices. At one extreme is the vice of excess, and at the other is the vice of deficiency.*

There are times in history when old ideas need to be scrapped in the light of new evidence. These times are both exciting and painful. Galileo Galilei experienced both sensations as he published the findings of his experiments. While his most original and inspired work — on the ways that gravity affects moving objects — was uncontentious, it was his investigations of Copernicus' concept that the Earth revolves around the Sun that got him into trouble. The problem was that the rulers of the day, the elite of the Church of Rome, had adopted Aristotle's ideas, and some felt threatened by change.

Galileo Galilei

1564–1642

Acquaintances
— Copernicus (1473–1543)
— Cardinal Roberto Bellarmino (1542–1621)
— Johannes Kepler (1571–1630)
— Benedetto Castelli (1578–1643)
— Viginio Cesarini (1596–1624)
— Vincenzo Viviani (1622–1703)

Galileo and motion

The difficulty with studying gravity was that it was too strong. If you dropped an object, it rushed to the ground before you had time to make any measurements. When Galileo arrived on the scientific scene the acquired wisdom was that when you released an object it accelerated to a particular speed and then got no faster. Whatever was pulling it was incapable of causing any more acceleration.

No one, however, was able to measure this to see if there was any truth in the concept. The limiting factor was the absence of a device that could quantify the passage of short intervals of time. At Padua in 1604, Galileo set out to investigate the issue. The first thing he needed was an instrument that could measure time, as he found that the contemporary method of counting heartbeats was too erratic to be useful.

To solve this, he devised a water clock, which he described in his book *Discourses on Two New Sciences* (1638): "For the measurement of time, we employed a large vessel of water placed in an elevated position; to the bottom of this vessel was soldered a pipe of small diameter giving a thin jet of water, which we collected in a small glass during the time of each descent... the water thus collected was weighed, after each descent, on a very accurate balance; the difference and ratios of these weights gave us the differences and ratios of the times..."

Even with an accurate way of measuring the passage of time, objects fell too fast to make any sensible measurements. Galileo's radical idea was to measure the rate that metal balls ran down a sloping wooden board that had a polished, parchment-lined groove cut in it to act as a guide. He correctly thought that gravity would act on the ball in proportion to the angle of the slope.

He then made a series of startling discoveries. Whatever slope he used he discovered a general rule: that the time for

above *Galileo's pendulum clock, c. 1642. It represents the first certain known attempt to apply a pendulum to control the rate of a weight- or spring-driven clock.*

the ball to travel along the first quarter of the track was the same as that required to complete the remaining three-quarters. The ball was constantly accelerating. Setting the principles for sound scientific method, he repeated his experiments—he called them *cimentos*, "ordeals"—hundreds of times, each time getting the same results.

His grounding in mathematics enabled him to spot the pattern—double the distance traveled, and the ball will be traveling four times faster; treble it and the ball will be moving nine times faster. The speed increases as a square of the distance. He also found that the size of the ball made no difference to the timing and that if the surface was horizontal, once a ball was pushed it would neither speed up nor slow down.

Scientific understanding would have to wait for Isaac Newton to make sense of these findings. His lack of complete comprehension, however, did not prevent him making accurate predictions about the way that gravity would affect objects thrown through the air. He argued that once an object like a cannon ball had been fired into the air, there would be nothing to slow it down. It would therefore continue at a constant horizontal motion, but would accelerate vertically toward the ground. This reasoning

above *One of the earliest refracting telescopes made by Galileo in 1610. It consists of a convex objective lens and a smaller concave eyepiece lens. The former gathers light from a distant object and bends it to form an upside-down image. The eyepiece lens clarifies this image by bending the light into parallel lines.*

led him to correctly predict that the path of a projectile was a parabola.

One rather nice myth was that he did his initial observations by dropping balls from the top of the leaning tower at Pisa, but while that is almost certainly untrue, another story is more likely to be grounded in reality. It appears that while sitting in Pisa cathedral one day he was distracted by a lantern that was swinging gently at the end of a chain. It seemed to swing with remarkable regularity, and when he started experimenting with pendulums he discovered that a pendulum takes the same amount of time to swing from side to side whether you give it a small push and it swings with a small amplitude, or a large push. The way to vary the rate of the swing is to either change the weight on the end of the arm, or alter the length of the supporting rope.

Galileo and temperature

According to his student Vincenzo Viviani, by the end of the 16th century Galileo had turned his attention to research on heat, and as part of this invented the thermoscope, which gave rise to the Florentine thermometer. Like modern thermometers Galileo's thermoscope operated on the principle that liquids expand when their temperatures increase.

Galileo and revolutions

If Galileo had stuck to monitoring heat and motion he would have contributed useful information to science, but it is unlikely that his name would be so readily recognized. Popular scandal is all too often more gripping than scientific endeavor, and scandal that involves religious authorities has a remarkable ability to make headlines.

By developing a two-lens telescope, Galileo opened a clearer window on the Moon, Sun, and planets. He was the first person to see sunspots, the first to spot the four main satellites of Jupiter and the first to record that the surface of the Moon had mountains and craters. So far there was nothing controversial.

He did, however, start to look into the work of the Polish-born medic, lawyer, and astronomer Nicolaus Copernicus, who in his 1543 landmark publication De revolutionibus orbium coelestium — roughly translated "On the Revolutions of the Heavenly Spheres" — presented his "heliocentric hypothesis." This said that the planets, including Earth, rotated around the Sun. The Earth was not the center of the universe.

Intrigued by the possibility, Galileo looked for evidence and found that Venus went through phases, much like the phases of the Moon. From this he deduced that Venus must be orbiting the Sun. For the Roman Catholic Church this presented two problems. First, it contradicted Aristotle's teaching that said all things rotated around the Earth; and it was also difficult to equate with some passages of the Bible which, if taken literally, again required the Sun to rotate around the Earth.

The tale of the 1633 Inquisition concluded with Galileo signing a document denouncing his own work. The following extracts serve to illustrate the nature of his "confession":

"That thou heldest as true the false doctrine taught by many, that the Sun was the center of the universe and immoveable, and that the Earth moved, and had also a diurnal motion: That on this same matter thou didst hold a correspondence with certain German mathematicians...."

"That the Sun is the center of the universe and doth not move from his place is a proposition absurd and false in philosophy, and formerly heretical; being expressly contrary of Holy Writ. That the Earth is not the center of the universe nor immoveable, but that it moves, even with a diurnal motion, is likewise a proposition absurd and false in philosophy, and considered in theology ad minus erroneous in faith..."

By this point Galileo was getting old and becoming blind. Confined by house arrest, he worked on his book Discourses upon Two New Sciences, which was smuggled out of Italy to a Dutch publisher. He died in his villa at Arcetri on January 8, 1642.

Timeline

1564
Born in Pisa

1581
Enters University of Pisa to study medicine and Aristotelian philosophy, but abandons medicine in favor of mathematics and physical science

1585
Leaves to study with Otilio Ricci in Florence

1589
Becomes professor of mathematics at the University of Pisa

1592
Becomes professor of mathematics in Padua

1610
Galileo is appointed chief mathematician to Cosmo II, the Grand Duke of Tuscany, a move that took him out of papal jurisdiction

1613
Writes to Father Castelli, suggesting that biblical interpretation be reconciled with the new findings of science

1615
Copy of Castelli's letter is handed to the Inquisition in Rome

1616
Galileo is warned by the Pope to stop his heretical teaching or face imprisonment

1632
Publishes Dialogues on the Two Chief Systems of the World, but the book is banned by the Catholic Church because it contradicts Aristotle and defends a Copernican understanding of the universe, a ban that is reversed only in 1822

1633
Is made to recant by the Inquisition and put under house arrest for the rest of his life

1637
Galileo becomes totally blind

1642
Dies at Arcetri near Florence on January 8

It could be straight from a novel. A story with a tragic start, but a glorious end. A sad child who hates his parents, adopts a reclusive lifestyle whenever possible, and prefers secrecy to publication, becomes one of the most famous scientists who ever graced the earth. If the novel were written, Isaac Newton would be the name of the central character. As a teenager he threatened to burn his parents' house, with them in it, but as an adult he was knighted for his work in mathematics and for establishing the nature of light and the universal force we now call gravity.

Isaac Newton

1642–1727

Acquaintances
— Leibniz (1646–1716)
— Edmond Halley (1656–1742)

An introvert genius

"I don't know what I may seem to the world, but as to myself, I seem to have been only like a boy playing on the sea-shore and diverting myself in now and then finding a smoother pebble or a prettier shell than ordinary, whilst the great ocean of truth lay all undiscovered before me." So wrote Isaac Newton about himself. Like many before and after he was passionate to search for truth, and determined to use every source of inspiration that he could lay his hands on. He scrutinized the Bible for clues, he pored over the works of the Greek philosophers, he joined forces with the clandestine alchemists, and he read the latest publications in the scientific press.

The results of these efforts led Albert Einstein to describe him as the one person who "combined the experimenter, the theorist, the mechanic and not least, the artist in exposition." In 1942 the English economist John Maynard Keynes called him the "last great mind which looked out on the visible and intellectual world with the same eyes as those who began to build our intellectual inheritance rather less than 10,000 years ago."

Tracking Newton's life is both sobering and inspiring—to ignore the pain, would do an injustice to his successes. His father died when he was just three months old and when he was two his mother remarried. Virtually abandoned, he was brought up by his grandparents. He was not an easy child and had little support. School reports describe him as "idle" and "inattentive." He lived a highly reclusive and secretive life, was terrified of criticism, and suffered at least two nervous breakdowns.

At Cambridge he aimed to study law and found the first two years dominated by the philosophy of Aristotle. In the third year, however, students were free to study other philosophers, and Newton turned to 16th- and 17th-century philosophers such as René Descartes, Pierre Gassendi, Thomas Hobbes, and Robert Boyle. In addition he read Nicolaus Copernicus' and Galileo Galilei's ideas of astronomy and Johan Kepler's theories about light. Newton claimed to have seen further than others "by standing on the shoulders of giants."

Defining differentiation; initiating integration

In 1663 Newton turned his interests to mathematics. He claimed to have been self-taught in geometry but went on to tackle one of the most critical outstanding problems of the time. Namely how to cope with curves.

The problem that Newton faced was that the angle of a curve, by definition, is constantly changing, so it was difficult to calculate at any particular point. Similarly, it was very hard to calculate the area under a curve. His solution was to develop the idea of Fluxions (from the Latin for "flow"), that have become known as differentiation and integration, where differentiation is a means of determining the slope of a line, and integration of finding the area beneath a curve. True to his passion for secrecy he kept this work hidden from all but his innermost circle of trusted correspondents, only publishing it in his book *Opticks* in 1704.

The colour of light

Since Aristotle, philosophers and scientists alike had believed that white light was a basic single entity. Newton, however, was not convinced. When building his telescopes he often found that the lens would generate colors. He argued that white light was a mixture of different types of rays, each of which is bent to a slightly different extent when it passes through a lens—each type of ray producing a different spectral color.

The work was controversial, because Newton explained the bending on the basis that light must be made of many different "particules." This was at odds with Hooke and Huygens, who were convinced that light was a wave. Given Newton's standing, science abandoned the wave theory for the best part of a couple of hundred years.

He was so convinced that lenses would always break light into its constituent "particles" that he devised a telescope that used a curved mirror instead of a lens. While we now know that high-quality lenses can keep all the colors of light together, this mistake in his thinking led to great improvements in the optical quality of telescopes. This advance is reflected in the fact that the world's largest array of telescopes is now named after him—the Isaac Newton Group.

An argument over his theory of color with English Jesuits living in Liège led to a violent exchange of letters that was closely followed by his first nervous breakdown. His mother died in the following year and he withdrew further into his shell, mixing as little as possible with people for a number of years.

Three universal laws of gravity

Newton, it is claimed, was sitting beneath an apple tree when, seeing an apple fall to the ground, he realized that the same force that pulled the apple also pulled the Moon. The story seems improbable, and is likely to be a myth, but the science is undoubtedly correct. At some time between 1665 and 1666 Newton had some flashes of true inspiration. He was well aware that Kepler had devised three laws of planetary motion. These were laws that explained how the planets moved, but were not expected to have any relevance to anything that happened on Earth.

Newton's genius was to recognize that any object that has mass will be pulled toward any other object. The larger the mass, the greater the pull. He devised his three laws of motion and said that these were true everywhere in the universe.

right *The 18th century English poet Alexander Pope expressed the enormity of Newton's work in his famous epitaph.*

Epitaph:
intended for Sir Isaac Newton
by Alexander Pope (1730)

*Nature, and Nature's laws
lay hid in night,*

*God said, Let Newton be!
and all was light.*

First law:

A body continues in a state of rest or uniform motion unless acted upon by an external force.

Second law:

Acceleration of a body is proportional to the force acting upon it and in line with that force.

Third law:

For every action there is an equal and opposite reaction.

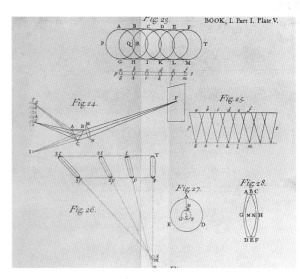

above *This plate from* Opticks *shows how a reflecting telescope works, how white light is dispersed into its constituent colors by a prism, and other optical principles.*

He published the concepts in *Principia*, a book now recognized as one of the greatest scientific books ever written. In it Newton analyzed the motion of orbiting bodies, projectiles, pendulums, and free-fall near the Earth. He further demonstrated that the planets were attracted toward the Sun by a force varying as the inverse square of the distance, and generalized that all heavenly bodies mutually attract one another.

Newton also proposed that, according to his laws, planets should follow elliptical, not circular paths, an idea that has now been thoroughly confirmed by observation. He also found inaccuracies in Kepler's work and corrected them.

Fascinated by alchemy, history, and religion

Various scientists have, besides their main area of study, a hobby that they dabble in. This was true of Newton. Strangely, though, his main concerns were not the areas for which he is most famous, but rather his passions for alchemy and the study of the influence of the Divine on the world. There is a mass of evidence that Newton was acquainted with clandestine circles of alchemists. In addition, he had a passionate belief in the Bible and was fascinated by the task of interpreting prophecy. His library contains vastly more books on areas now described as the humanities than on science.

Newton, however, brought his science into this area of work as well. It dawned on him that some people in recording history had commented on the position of various planets at times when events took place. They had done so because they believed the planets were influencing the events, but Newton realized that this gave him a natural clock against which he could order key moments in history. Using this theory he sought to reconcile Greek mythology with the Bible, and tried to make Jewish and pagan dates compatible.

Avidly reading works by the Fathers of the Church, he came to the conclusion that he disagreed with aspects of the Church's central creed, but he went to his death still expressing a strong belief that God had an ongoing involvement in the world and in Nature.

Timeline

1642
Born at Woolsthorpe, near Grantham in Lincolnshire. There is something vaguely poetic in the fact that Isaac Newton was born in the year that Galileo Galilei died. It is as if a baton changed hands, passing the challenge of elucidating the nature of gravity to a new generation

1661
Enters Cambridge and helps pay for his education by being a servant to other students

1667
Elected a fellow of Trinity College

1665–1666
Because of the Plague, Cambridge University was closed and Newton moves to Lincolnshire. Undisturbed by other people, these are the most productive months of his life — "the prime of my age for invention"

1669
Becomes Lucasian Professor of mathematics. In the same year he moves to London as Warden of the Royal Mint

1671
Elected as a fellow of the Royal Society

1678
Suffers nervous breakdown

1693
Suffers second nervous breakdown

1703
Newton becomes president of the Royal Society

1704
Publishes *Opticks*

1705
Knighted in Cambridge

1727
Dies in London on March 20

Michael Faraday

1791–1867

Acquaintances
— John Tatum (1772–1858)
— Alessandro Giuseppe Anastasio deVolta (1745–1827)
— Humphry Davy (1778–1829)
— Jean Jacques Ampère (1800–1864)

While Dick Whittington famously set off to London to make fame and fortune, Michael Faraday's father headed to the metropolis in the hope of making ends meet. In so doing he gave his son the opportunity of growing up in a rich environment, where anyone, if they wanted to, could hear about science. Perseverance and good fortune elevated Faraday to a senior position within English science, enabling him to invent equipment and conduct experiments that introduced humankind to the concept of electricity on demand. His public lectures enthralled hundreds of people by giving them a glimpse into the world of science.

A poor beginning

Life as a blacksmith was tough and in 1791 James Faraday and his wife, Margaret, moved south from Yorkshire to see if they could eke out a living in Newington Butts. At that time it was a village on the edge of London, an area that is now right in the center of the heaving metropolis. In 1795 they moved into the center of the bustling city, little knowing that this would give their son an incredible chance to change the world.

With poor parents, Michael Faraday had only a very basic education and throughout life remained mathematically almost illiterate. In retrospect, his biggest break came when he went to work as an apprentice bookbinder, and his workplace proved to be a rich environment for an inquisitive mind. People with ideas wrote them down and discussed them in books, so the best way of keeping up with the latest thinking was to read the most recent books. Better still would be to read them before they were even published. Soon Faraday was reading and taking notes from the books that passed through his hands, giving him unprecedented access to emerging knowledge. His family belonged to the Sandemanians, a religious group that believed in a literal understanding of the Bible, and in his religious reading he picked up *The Improvement of Man* by the preacher Isaac Watts, which inspired him to read more and attend lectures.

Through perseverance and good luck he found himself working for Humphry Davy at the Royal Institution, and later was given a flat at that prestigious London address, just off Bond Street. Davy was fascinated by chemistry, but also determined to draw the attention of as many people as possible to the wonders of science.

Electrical conclusions

Fascinated by chemistry, Faraday must have felt like a child let loose in a sweet shop when he entered the Royal Institution. However, his most important work was on electricity in general and electromagnetism in particular.

In October 1821 he published a paper in the *Quarterly Journal of Science* entitled "Some New Electro-Magnetical Motions and on the Theory of Magnetism." This was groundbreaking research. The paper describes the first conversion of electrical energy into mechanical energy and vice versa. Faraday had taken a magnet and found that it would rotate around a fixed wire if you passed a current through the wire. In effect, you could get mechanical energy in the form of a moving magnet from electrical energy passing through the wire. It was the forerunner of the electric motor.

That wasn't all. Faraday also noted that if you fixed the magnet in place and let the wire move, it would rotate around the magnet when you passed a current through.

Genius is not always in the initial observation, but in the interpretation of that observation. Faraday's belief was that there must be circular lines of force radiating from the wire that interacted with the magnet. It was an idea that went on to be proved and embellished as "field theory" by future physicists, including James Clerk Maxwell who developed mathematical explanations of the findings. At the time, however, his interpretation was rejected by most mathematical physicists of Europe, since they assumed that electric charges attract and repel each other by actions at a distance, making such lines unnecessary.

A decade later, Faraday revealed his next breakthrough. He had taken an iron ring and wrapped a length of insulated wire around one side of it, and then wrapped a second insulated wire around the other side of the ring. He anticipated that if you passed a current through the first

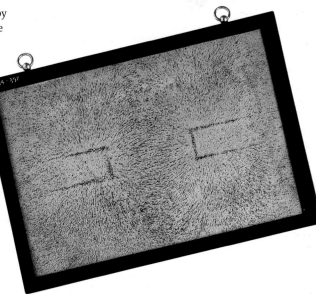

above *Lines of magnetic force prepared by Faraday in 1852. This "delineation of lines of magnetic force by iron filings" was produced by sprinkling iron filings over a pair of magnets and chemically fixing the pattern produced to the paper.*

wire it would establish a field in the ring that would induce a current in the second wire. At first it seemed as if the experiment was a failure. No current was induced. But then he made a critical observation. Each time he switched the current on and off, the needle on the galvanometer, which was measuring the current in the second coil, flickered. It was the forerunner of a transformer.

Thinking about this, he realized that whenever the current in the first coil changed, current was induced in the second. To test this concept he slid a magnet in and out of a coil of wire. While the magnet was moving the galvanometer needle registered a current. As he pushed it in it moved one way, as he pulled it out the needle moved in the opposite direction. Next, he set up a copper disk and caused it to spin in a gap between the poles of a magnet — this time he got a constantly flowing electrical current. What he had in fact made was the world's first dynamo. Faraday went on to realize that static electricity, the electricity in a battery, and the charge given out by electric eels were all the same thing.

Electrolysis

Combining his pursuit of chemistry and electricity, Faraday made some critical discoveries about the effects of passing electrical currents through chemical solutions. His understanding of what occurred was so different from previously held views that in 1834 he published a paper in the *Philosophical Transactions* of the Royal Society that introduced six new words to the realm of science — words which, when carefully defined, enabled him to describe what his work was all about.

Instead of calling the object that scientists dangled into a solution a "pole," Faraday introduced the word electrode. The electrode where negatively charged gases like oxygen and chlorine collect when a current is passed through a solution he called an anode, and gave the name cathode to the electrode where positively charged chemicals collect. Anything released at either electrode he called an electrolyte. These are divided into anions that collect at the anode and cations, which are drawn to the cathode.

If the proof of the pudding is in the eating, the proof of new words is in their usefulness and longevity. Faraday would no doubt be thrilled to know that these words are still in use today.

Public demonstrations

Throughout his life, Faraday never forgot the thrill he had received from attending public lectures and, once in a position of knowledge and influence, he was keen to join in the tradition.

In 1827 he gave the first Christmas Lecture in the Royal Institution, a lecture that was based around 86 demonstrations of various chemical principles. It was a mixture of pyrotechnics, thunder flashes, and color changes. At one point he demonstrated how chemicals combine and produce new products by mixing saltpeter, crushed sulfur, and powdered charcoal — the standard recipe for gunpowder. When ignited, he showed that it produced sound, light, and heat, a few gases, and very little remaining residue.

At another point in the lecture he showed how to weigh air, by weighing a glass globe, then pumping the air out and weighing it again. It was now lighter. The difference was the weight of air it had contained. Even his rudimentary mathematics was sufficient to calculate that this indicated there were about two tons of air in the theater.

Private religious belief

Throughout his life Faraday maintained his Sandemanian belief. Part of the lifestyle that resulted from it was that he refused to hoard or save wealth. One consequence was that when he married he had no money for accommodation — a need that was met by the Royal Institution — but again on retirement he was once more stuck. This time Queen Victoria stepped in by offering him accommodation in Grace and Favor apartments at Hampton Court, where he lived with his wife until he died.

Timeline

1791
Born (September 22) in Newington Butts, Surrey (now London)
1795
Moved to Jacob's Well Mews, London
1805
After a year as an errand-boy, Faraday is taken on by bookseller George Riebau as an apprentice bookbinder
1810
Starts attending lectures at the house of John Tatum
1812
Attends lectures at the Royal Institution given by Humphry Davy. He wrote to Davy asking for a job, but Davy recommended he keep to bookbinding. In this year his apprenticeship ended and Faraday got a job as a bookbinder
1813
Davy temporarily blinds himself in an experiment and invites Faraday to a short-term clerical post. Faraday accepts and months later replaces Davy's technician, who is sacked for fighting
1813
Faraday joins Davy on an 18-month European tour in which he meets Ampère and Volta
1821
Marries Sarah Barnard and the couple move into an apartment in the Royal Institution
1821–1831
Faraday works on chemistry
1831–1839
Works on electricity
1839
Faraday suffers a breakdown and never regains his full physical or intellectual strength
1867
Dies on August 25 in Hampton Court, Middlesex

Given the amount of life that is now dominated by, or at least organized around, radio and television, it's difficult to imagine life without radio waves. In fact there never was a time when this electromagnetic radiation did not exist, but it took a combination of a mathematician and a physicist to pin them down so that they could be studied and measured. When Heinrich Hertz sent the first pulses of radiation across his laboratory he had no idea of the effect that it would have on human history—and, dying young, he only ever caught a glimpse of the potential.

Heinrich Hertz

1857–1894

Acquaintances
— Hermann von Helmholtz (1821–1894)
— Gustav Robert Kirchoff (1824–1887)

Revealing radio

When Heinrich Hertz started his experimental work at the University of Bonn he was well aware of the pioneering thinking that had come from British scientist James Clerk Maxwell. Maxwell had produced a series of mathematical equations that predicted the existence of electromagnetic waves. The problem was that no one had found a way of deliberately creating them.

In 1887 all that changed. Hertz, who as a child had enjoyed building things, set up an oscillator made of polished brass balls, each connected to an induction coil. These balls were separated by a tiny gap and, when Hertz applied a current to the coils, sparks would leap across the gap. It was an interesting demonstration, but nothing particularly new at the time. Hertz, however, reasoned that if Maxwell's predictions were correct, then each spark would emit electromagnetic waves that should radiate through the laboratory.

To test this, he made a simple receiver. It consisted of a loop of wire. At the ends of the loop were two more small balls, once again separated by a tiny gap. This receiver was placed several yards from the oscillator.

According to theory, if electromagnetic waves were spreading from the oscillator sparks, they would induce a current in the loop that would send sparks across the gap. This happened, producing the first transmission and reception of electromagnetic waves. It would take a few years to refine this into a device that had the potential of transmitting a message, and that feat was done by Italian electrical engineer Marchese Guglielmo Marconi who built the first radio equipment. With this, he could send a signal from his transmitter and pick it up by a receiver placed 10 yards away. The receiver simply rang a bell, but at that time in history, this was progress.

During this research, Hertz accidentally discovered the photoelectric effect whereby light falling on special surfaces can generate electricity.

Hertz didn't stop when he had achieved his first results. He soon spotted that electrical conductors reflect this electromagnetic radiation and that non-conductors allow most of the waves to pass through. In addition he found that they could be focused by concave reflectors. This radiation was behaving with all the properties that would be expected for waves. Moreover, he showed that the nature of their vibration and the susceptibility to reflection and refraction were the same as those of light and heat waves. At the time this was stunning, though now we take it for granted that electromagnetic radiation includes radio waves, microwaves, light, and infrared heat. In recognition of his work, the unit of frequency of a radio wave, one cycle per second, is called a hertz.

Reluctant hero

It was not that he had invented anything. Hertz had simply introduced humanity to the electromagnetic radiation that had been there all along, just waiting for someone to recognize it. In fact, Hertz seems to have been rather nonchalant about the discovery at first. There is a story that

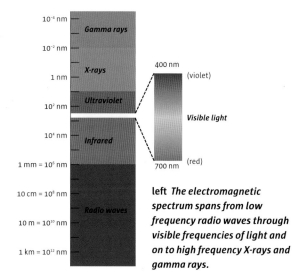

left *The electromagnetic spectrum spans from low frequency radio waves through visible frequencies of light and on to high frequency X-rays and gamma rays.*

after demonstrating his groundbreaking experiment, one of his students asked him what use it might be, to which he apparently replied: "Of no use what-so-ever... This is just an experiment that proves Maestro Maxwell was right, we just have these mysterious electromagnetic waves that we cannot see with the naked eye." "So what next?" inquired the student. "Nothing, I guess," was the reply.

But if Hertz was slow to realize the implications, others were quick to pick it up. In 1891 the English mathematical physicist Sir Oliver Heaviside said: "Three years ago, electromagnetic waves were nowhere. Shortly afterward, they were everywhere."

Timeline

1857
Born on February 22 in Hamburg, Germany, to a Jewish father, a prominent lawyer, who had converted to Christianity, and received a comprehensive education in the humanities and natural sciences
1880
PhD from the University of Berlin
Works as assistant to the eminent physicist Hermann von Helmholtz
1883
Takes a post as lecturer in physics at the University of Kiel
1885
Becomes professor of experimental physics at the Karlsruhe Polytechnic at the age of 28
1886
Marries Elizabeth Doll, daughter of a Karlsruhe Polytechnic professor
1889
Appointed professor of physics at the University of Bonn
1894
On January 1 dies of blood poisoning at the age of 37 after a long illness

By the late 1800s the science of thermodynamics was developing to the point that people were beginning to understand the nature of energy. The traditional view was that any amount of energy could be infinitely divided into smaller and smaller "lumps." Planck's work on the laws of thermodynamics and black-body radiation led him to abandon this classical notion of the dynamic principles of energy and formulate the quantum theory, which assumes that energy changes take place in distinct packages, or quanta, that cannot be subdivided. This successfully accounted for certain phenomena that Newtonian theory couldn't explain.

Max Planck

1858–1947

Acquaintances
— Thomas Andrews (1813–1885)
— James Prescott Joule (1818–1889)
— Hermann Ledwig Ferdinand von Helmholtz (1821–1894)
— Gustav Robert Kirchoff (1824–1887)
— James Clerk Maxwell (1831–1879)
— Ludwig Boltzmann (1844–1906)
— Marie Curie (1867–1934)
— Albert Einstein (1879–1955)
— Niels Bohr (1885–1962)

Two theories of dynamics

As the 19th century drew to an end, many physicists were asking searching questions of time-honored Newtonian mechanics. In particular, did it still describe all of nature? In answering the question scientists started to group into two camps. One was looking for an answer by studying what they called "electrodynamics," the relationship between mechanics and electricity. The other was looking into thermodynamics and its two basic laws. The first law recognized that energy could not be made or destroyed, but was always conserved, and the second law was drawn from an understanding that heat won't pass from a colder to a hotter body.

The study of thermodynamics was based on the assumption that matter was ultimately composed of particles. However, this posed a problem since atoms had not been discovered. Instead, the conventional view was that matter was continuous, not made up of discrete building blocks. Starting in the 1870s, Ludwig Boltzmann had proposed an explanation of thermodynamics, saying that the energy contained in a system is the collective result of the movements of many tiny molecules rattling around. He believed that the second law was valid only in a statistical sense, it only worked if you added up all the bits of energy in all the little particles.

Boltzmann had supporters, but there were many who doubted. Among the detractors was one Max Karl Ernst Ludwig Planck. He was fascinated by the second law of thermodynamics, but rejected Boltzmann's statistical version because he doubted the atomic hypothesis on which it rested. In 1882 he falsely stated: "in spite of the great successes of the atomistic theory in the past, we will finally have to give it up and to decide in favor of the assumption of continuous matter."

During the 1890s Planck began to see that the atom hypothesis had the potential of unifying many different physical and chemical phenomena, but he was still skeptical, and his own research was aimed at finding a non-atomic solution.

From black-bodies to quanta

Planck and his contemporaries looked to Scottish physicist James Clerk Maxwell's theories of electrodynamics to find the answers, but were unsuccessful. Instead, a new understanding began to emerge when they turned their attention to black-body radiation. A black-body is a

above *Optical image of a so-called "runaway star." This star moves at about 80 kilometres per second through the gaseous inter-stellar medium, fast enough to create a pronounced bow shock front (magenta).*

theoretical object that absorbs all of the radiation that hits it. Because it reflects nothing it will be black. While a black body does not reflect radiation, it still radiates heat itself. Otherwise it would just keep on absorbing and its temperature would rise indefinitely. The closest thing in existence to it now is the US spy plane the Blackbird, which is coated with an absorbent pigment that attempts to absorb all radiation.

The first person to start thinking about black-bodies had been Planck's predecessor as professor of physics in Berlin, Robert Kirchhoff, who argued that such radiation was of a fundamental nature. By the 1890s several physicists were investigating the spectral distribution of the radiation. In 1896 Wilhelm Wien produced a radiation law that agreed with experimental observations but that, according to Planck, was theoretically unsound, so he rejected it. In 1899 Planck produced a new version that incorporated some of Boltzmann's ideas which is sometimes referred to as the Wien-Planck law. Planck was satisfied. At that point he felt that the law agreed with experimental data and had a sound theoretical basis.

Sadly for Planck, it turned into a beautiful theory destroyed by ugly facts. Experiments performed in Berlin showed that it did not work for low frequency radiation. After a revision of his ideas he came up with a new concept that included a value for a constant that he termed "b," and he presented it at a meeting of the German Physical Society on October 19, 1900. The new theory, however, still did not take into account any notion of particles or quanta of energy. With the benefit of hindsight we can see that the true answer was staring Planck in the face, but he was so certain of the continuous nature of matter that he couldn't see it for himself.

$$P_\lambda = \frac{2\pi hc^2}{\lambda^5(e^{(hc/\lambda kt)}-1)}$$

P_λ=Power per m^2 area per m wavelength
h = Planck's constant (6.626 x 10^{-34} Js)
c = Speed of Light (3 x 10^8 m/s)
λ = Wavelength (m)
k = Boltzmann Constant (1.38 x 10^{-23} J/K)
T = Temperature (K)

above *While this equation is complicated, physicists can use it to study the thermal radiation emitted from any object with a temperature above absolute zero.*

Then two months later, and "as an act of despair," he renounced classical physics and embraced quanta. The final straw had been a concept developed by John Rayleigh and James Jeans that became known as the "ultraviolet catastrophe" theory. In June 1900 Rayleigh pointed out that classical mechanics, when applied to the oscillators of a black-body, leads to an energy distribution that increases in proportion to the square of the frequency. This conflicted with all known data.

Planck's desperation led him to introduce what he called "energy elements" or quanta. In his presentation to the German Physical Society on December 14, 1900, Planck said that energy was "made up of a completely determinate number of finite equal parts, and for this purpose I use the constant of nature h = 6.55 x 10^{-27} (erg sec)."

Quantum theory was born, though it would take another two or three decades and a few more talented minds to realize the implications of the new era.

above *Niels Bohr (left) with the German physicist Max Planck. Both were central figures in the development of quantum theory in the early 20th century.*

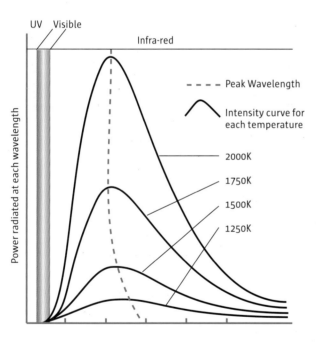

above *Black body radiation curves are affected by the temperature of the emitting object. In this graph, all the temperatures yield radiation in the infrared part of the spectrum.*

Timeline

1858
Born in Kiel, Germany, into an academic family. His father was the professor of constitutional law at Kiel, and both his grandfather and great-grandfather had been professor of theology at Göttingen

1867
Moves to Munich, where he went to school

1874
Studies at Munich under Gustav Kirchhoff. Before starting he discussed the research prospect in physics with the professor of physics Philipp von Jolly, who is supposed to have told him that physics was essentially a complete science with little prospect of further developments. Thankfully he appears to have ignored the comments, before moving to Berlin to study under Hermann von Helmholtz

1889–1926
Professor of physics, Berlin

1900
Announces his Quantum Theory

1914–1918
Eldest son dies in the First World War

1918
Awarded the Nobel Prize

1926
Elected to Foreign membership of the Royal Society

1944
Youngest son, Erwin, is executed when declared guilty of being involved in a plot to assassinate Hitler

1947
Dies at Göttingen on October 3. After hearing about his death Albert Einstein wrote: "How different and how much better it would be for mankind if there were more like him... It seems that fine characters in every age and continent must remain apart from the world, unable to influence events."

There are no absolutes—everything is relative, except that is, for the speed of light. That alone is always constant. Measurements of speed, mass, space, and time are all dependent on who is measuring them and what they are doing at the time. Space is curved and all the energy locked in the atoms of one kilogram of mud is the same as the explosive power of 30 million tons of TNT. To say that Einstein stirred things up would be a monumental understatement—he revolutionized humanity's view of the Universe.

Albert Einstein

1879–1955

Acquaintances
— Hendrik Antoon Lorentz (1853–1928)
— Max Planck (1858–1947)
— Paul Ehrenfest (1880–1933)
— Max Born (1882–1970)
— Niels Bohr (1885–1962)
— Erwin Schrödinger (1887–1961)
— Louis Victor de Broglie (1892–1987)
— Werner Heisenberg (1901–1976)

1905, as we like to measure it

Imagine working as an editor at an academic journal, the Annalen der Physik. It is March 1905 and you open an envelope and tip out a handwritten draft of a paper, with a note asking for it to be published "if there is room," signed by an unknown person calling himself Albert Einstein.

The paper elegantly updates Max Planck's theory of radiation. Two months later another letter arrives. This time the paper defines the way that gas particles bounce off each other. June arrives, and so does a third paper. This one makes the rest look simple, in that it purports to modify the theory of space and time. You would surely spend some time trying to work out whether these were hoaxes, coming from a crackpot, or the work of a genius. History tells us that the latter was true.

Einstein had a problem—he was troubled by light. By the beginning of the 20th century scientists were certain that light traveled at 300,000 kilometers per second. They had given the value the name "c." So far so good. But they stumbled when they tried to work out how light, being a wave, was carried along. They knew that sound waves need air and water waves need water, but were searching for the material that supported light. The general feeling was that space was filled with ether, and light traveled through this ether. For the best part of 40 years, scientists searched for ether to no avail.

In 1887, two American scientists, Albert Abraham Michelson and Edward Williams Morley had set out to detect ether by using the Earth's movement through space. They assumed that the ether in space would be stationary and that they could measure it rushing past the Earth— much like feeling the wind in your face while riding a bike. Their plan was to measure the speed of light in the direction that the Earth was moving, and compare it with the speed of light traveling in the opposite direction. Given that they believed that light moved at a constant speed through ether, the difference should show how fast the Earth was moving. To their surprise, whatever way they looked, light always traveled at the same speed.

Working in Bern, Switzerland, Einstein also started to question whether ether actually exists. He asked the

question, what would it be like to ride on a light wave—to travel at the speed of light? If you held a mirror in front of yourself, would the light ever make it into the mirror and reflect back? Would you see your reflection, or would the mirror always appear to be blank? If so, relative to the traveler, light would appear to have stopped. For 10 years, from 1895–1905, Einstein puzzled over this, because he was convinced that somehow, his image should not disappear.

In order to resolve this situation, Einstein started looking at issues initially raised by Galileo Galilei, who had pointed out how difficult it was to measure the true speed of an object. The problem was knowing whether you and the planet that you were sitting on were also moving. When we look at objects at rest on a table, they are in fact hurtling through space. They appear to be at rest because we and the table are moving at the same speed. This principle of relativity said that you are not able to tell that we are moving without some other external frame of reference.

Einstein started to consider the consequences for light. He argued that if your image disappeared in the mirror when you sat on your hypothetical light beam, then this would be evidence that you and your frame of reference were moving. But he had just agreed that you could never find evidence of your own movement without looking outside your frame of reference. It would violate the principle of relativity, and Einstein wasn't prepared to do that.

The mind-bending moment came when Einstein made a shocking suggestion. Time was not constant. Worse than that, time varied according to how fast the observer was traveling. The traveler moving at just less than the speed of light would still see a normal image in the mirror, because time for that person would slow down.

It's a wacky concept, but you can measure the effect. Synchronize two atomic clocks. Place one in a supersonic jet and leave the other on the runway. Fly the jet around for a bit, return to the ground and compare the two. The clock that has lived at high speed has experienced less time. It will not have counted as far. The experiment works—it has been done a number of times. The time interval between two events is not an absolute quantity, but it depends on the velocity of the person who is measuring the interval.

If you ever managed to travel at the speed of light, time would stop, although you would be unaware of it. Traveling at greater than the speed of light, said Einstein, is therefore impossible—though a few people have questioned whether it might allow you to set the clock running backwards. This became known as the theory of special relativity, though at the time it received little if any acclaim.

E=mc²

It was another aspect of Einstein's theory that was unexpected even to the man himself, that did have a very real impact on the world. Special relativity led to the realization that $E=mc^2$.

In 1905 two key principles underpinned physics and chemistry. First, in any situation of change, such as a chemical reaction, if you weigh how much material you have at the beginning you will always end up with the same

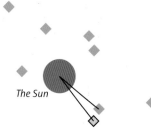

The Sun

above *On May 29, 1919 English explorer Arthur Eddington photographed an eclipse of the Sun which proved Einstein's theory that light bends when it moves past a massive object.*

amount at the end. Mass is always conserved. Secondly, if you measure all the energy in a system at the beginning of an experiment, you will always have the same amount at the end, even if a lot of it is present in the form of heat.

Einstein connected these two concepts and caused an explosion in thought and reality. He said that the mass of an object represents the energy locked up in it. If, however, you make the object travel fast, then the energy it contains is the sum of mass, plus the energy needed to make it move. This added energy must be represented as added mass. The mass of an object will increase as it gets faster.

Simple arithmetic led him to conclude that the extent of that increase in mass was once more related to the speed of light. There was something remarkably constant in that number. Calculations indicate that if you were to travel at 40% of the speed of light you would get 10% heavier. At 99% of c your mass will increase sevenfold and at 99.99% of c your mass will rise to 70 times your stationary mass. It is impossible to get to 100% of c, because at that point you would be infinitely heavy, and that clearly is not possible.

If mass is related to energy, then it should be possible to destroy the mass and release the energy. Hiroshima and Nagasaki went down in history as the first places to see the consequence of that theory unleashed in anger. The paradox is that Einstein, the ardent pacifist, had revealed an aspect of nature that others took and used to create atomic weapons.

General relativity

After Einstein moved to Zurich in 1909 he started developing his ideas. He wanted to take the theory of special relativity and apply it more broadly—to come up with a general theory. He turned his attention to gravity.

Isaac Newton had shown that two masses attract each other. The attractive force was gravity. Einstein adapted this by realizing that gravity was equivalent to acceleration. In a feat of mental gymnastics he also showed that gravitational fields distort time and space.

In 1915, he published his ideas that a light beam could be bent by gravity. This was not because light has weight—calculations based on that theory get the amount of bending wrong by a factor of two. Rather, Einstein speculated that the gravity of a massive body could curve the "spacetime" that surrounded it. Light, passing through this curved spacetime, would follow the straightest path that it could, and would consequently follow a curved path.

Einstein had used his concept that light would be affected by gravity to suggest that if light passed near to the Sun its path would bend. If the light was en route from a star to Earth, the effect of this to an Earth-bound observer would appear as if the star were in a different position. In the year after the end of World War I, on May 29, the Moon passed in front of the Sun. This momentarily blocked out the blazing light and let astronomers take photographs of the stars that appeared to be right next to the Sun. Amazingly the dots of light from these stars were in the wrong place. The light had indeed been deflected by the Sun, and Einstein was launched in the world's press as an international superstar genius.

Timeline

1879
Born 14 March, in Ulm, Germany
Spends his early years in Munich

1894
After repeated failures of the family's electrical engineering business, the family moves to Milan, Italy

1895
Fails an entrance exam for the Swiss Federal Institute of Technology, and goes to school in Arrau, Switzerland, but misses classes and prefers reading about physics or playing his violin

1896
Renounces German citizenship

1900
Leaves school, but his teachers don't recommend him for a university course

1901
Takes Swiss citizenship and publishes a paper about the forces between molecules

1902
Secures a position as an examiner in a Swiss patent office in Bern

1903
Marries Mileva Maric, a Serb whom he had met in one of his physics classes, with whom he had two sons in 1904 and 1910, before later separating in 1914 and getting divorced in 1919

1905
Receives his doctorate from the University of Zurich

1905
Publishes three papers on theoretical physics, the third of which is entitled "On the Electrodynamics of Moving Bodies," and contains the "special theory of relativity"

1909
Gets first academic appointment at the University of Zurich

1911
Moves to the German-speaking university in Prague as professor of theoretical physics

1913
Appointed director of the Kaiser Wilhelm Institute for Physics

1916
Proposes the general theory of relativity, which is proved correct three years later

1919
Marries Elsa Löwenthal, a cousin
Eddington's report of starlight bending around the sun during a solar eclipse confirms Einstein's theory

1922
Nobel Prize for Physics

1933
Emigrates to Princeton, New Jersey, USA

1939
Einstein joins other scientists in writing to President Franklin D. Roosevelt, saying that it is possible to build an atomic bomb, and suggests that Germany may already have the technology

1952
Offered and declined presidency of Israel

1955
Dies in his sleep in Princeton, USA, on April 18

The general public had problems coming to terms with Einstein's theories, but were still captivated by this enigmatic character and drawn to realize that he must be making some sort of sense when they saw physical manifestations of his work in moving stars and exploding bombs. Relativity, however, proved to be the easy bit, because as physicists looked harder into gravity and electrodynamics they decided that the three dimensions of space that we were used to were not enough. Somewhere hidden from sight there must be more. Theodor Kaluza and Oskar Klein made their fame among scientists by showing that recognizing the existence of a fifth dimension could solve their problems.

Theodor Kaluza & Oskar Klein

Unification comes in higher dimensions

Reality is more complex than it might at first appear. We are used to the idea that an infinitely small dot has no dimensions, that a line has one dimension, a flat surface has two and that a solid object has three dimensions. We can draw those dimensions on a graph and give them coordinates using x, y and z axes. If you want to describe how that object alters through history, you need to add a fourth dimension — time.

That much is pretty obvious, and you can easily measure all of the dimensions with a ruler and clock. They underpin much of basic science that at root devotes itself to observing, counting, and measuring.

The problem was that the more carefully theoretical physicists looked at Einstein's general theory of relativity and Maxwell's theory of light, the more they were convinced that there must be a way of joining them together — of creating a unified theory. The quest started in the first half of the 20th century and this Holy Grail of theoretical physics is still proving elusive at the opening of the 21st.

Two people, Theodor Kaluza and Oskar Klein, did however open up a new era in thought when they independently came to the same mind-bending conclusion. You could draw the two theories together so long as you introduced a fifth dimension. If you shake your head and think that this is mad, you are not alone. Even Einstein, who was quite used to stirring up intellectual debate, rejected the idea when it was first presented to him, before embracing the concept and taking trouble to publicize it and help the proponents to gain promotion within their institutions.

So where is the fifth dimension and why can't we see it? The answer, claim physicists, is that it is very small and curled up in a circle. Klein suggested that a particle moving a short distance along this fifth axis would return to where it began. They insist that it is more than an aberration of a mathematical equation, that it has a physical existence, and that the distance traveled before a particle gets back to its starting place, the circumference of the circle, is the size of this fifth dimension. All electromagnetic waves can be thought of as vibrations of this fifth dimension.

Stringing it together

For many years the Kaluza-Klein theory was more or less a curiosity, and physicists busied themselves working on more standard models of particles. It has however found a new dawn in the light of string theory.

As its name hints, string theory proposes that everything is made of strings. Again, you need to be prepared to stretch your mind. Strings are the smallest possible particles, with a length of 10^{-33} cm, no width, and no height. They can be open with two ends, or closed into a circle. A string has a fixed point in time and space, and you can chart its movement on a space-time graph.

Intriguingly, string theory only works mathematically in equations that use 10 dimensions. The math fails with any other options. Kaluza-Klein theory explains the existence of the six dimensions that we cannot ordinarily experience. They are, apparently, all curled up in extremely tiny balls.

M is for mother

The search continues for ways of making sense of all the mathematics and observations that are coming from the enormous particle accelerators that have been built in Europe and America. At one point there were five separate string theories, but in 1994 they were unified into a single M-theory. This bizarrely requires 11 dimensions, but it still is not fully developed.

Timeline

1885
Theodor Kaluza is born in Ratibor, Germany, on November 9
1894
Oskar Klein is born in Mörby, Sweden
1917
Klein works with Niels Bohr in Copenhagen
1919
Kaluza writes to Einstein explaining his ideas about how to unify Einstein's theory of relativity with Maxwell's theory of light
1921
Einstein eventually sees the potential of Kaluza's work and agrees to publish his theory
1923
Klein marries Gerda Koch and moves to the University of Michigan. While teaching electromagnetism he begins to consider introducing a fifth dimension in order to unify gravitational and elecromagnetic fields
1925
Klein returns to Copenhagen and contracts hepatitis
1926
Klein becomes aware of Kaluza's work, and publishes his own ideas in the science journal *Nature*
1935
Kaluza is made a professor, after Einstein writes to Kiel University to explain his brilliance
1945
Kaluza dies in Göttengen, Germany, on January 19
1977
Oskar Klein dies in Stockholm on February 5

Theodor Kaluza	Acquaintances
1885–1954	—Albert Einstein (1879–1955)
	—Erwin Schrödinger (1887–1961)
Oskar Klein	—Werner Heisenberg (1901–1976)
1894–1977	—Niels Bohr (1922–1962)

Erwin Schrödinger

1887–1961

Acquaintances
—Arnold Sommerfeld (1868–1951)
—Albert Einstein (1879–1955)
—Hermann Weyl (1885–1955)
—Hans Reichenbach (1891–1953)
—Louis de Broglie (1892–1987)
—Werner Heisenberg (1901–1976)
—Paul Dirac (1902–1984)

Quantum theory placed a bomb under the worldview of classical physics and in the end overthrew it. One of the critical steps in this rebellion was achieved when Erwin Schrödinger formulated his theory of wave mechanics, in which he suggested that an electron in an atom behaves like a wave. He was driven by beauty, by his underlying principle that if a solution was not mathematically beautiful then it was almost certainly incorrect. Schrödinger's work received vital stimulation when he read Louis de Broglie's Ph.D. thesis, and was officially recognized when Schrödinger shared the 1933 Nobel Prize for Physics with Paul Dirac.

Electrons wave hello

In 1900 Max Planck had first suggested that energy came in lumps. This led to the idea that light—which is a form of energy—was also composed of particles. It seemed unlikely at first, but Einstein had developed the concept to the point where it had considerable credibility, and light particles became known as photons.

While light was clearly a particle it also had wave-like properties. Planck's work had shown that different light came in different colors because the photons had different amounts of energy. If, however, you divided that energy by the frequency at which that color of light was known to oscillate, you always ended up with the same value, the so-called Planck's constant.

So much for light. But what about particles of matter? Maybe they also had wave properties? The question started to find an answer when Louis de Broglie, a 20th-century aristocratic French physicist, suggested that particles of material looked like localized lumps because we were not able to monitor them closely enough. More sensitive observations, he believed, would reveal that they too had wave properties.

Drawing support for his ideas from Einstein's theory of relativity, de Broglie showed that using Einstein's equations, he could represent motion of matter as waves. He presented his findings in his 1924 doctoral thesis *Recherches sur la Théorie des Quanta* (Researches on the quantum theory). This was confirmed by experimental work with electrons carried out by American physicists Clinton Joseph Davisson and Lester Hallbert Germer in 1927, which showed that electrons, while being particles, also behaved as waves. Planck had changed our view of light, now de Broglie had challenged matter.

Schrödinger's part in the continuing revelation was to take de Broglie's observations and develop a wave equation that described the way that electrons behaved. He used the equation to define the ways that electrons could move within atoms, and discovered that the equations only worked when their energy component was held at a multiple of Planck's constant.

In 1933 Schrödinger picked up the Nobel Prize for Physics, but as he did so he paid tribute to Fritz Hasenhörl, the physics teacher who had stimulated his imagination when he was a student at the University of Vienna. Hasenhörl had been killed in World War I, but during his acceptance speech, Schrodinger remarked that but for the war it would have been Hasenhörl, not him, receiving the honor.

A positive partner

When Schrödinger stepped onto the stage to collect his Nobel medal he was accompanied by Paul Dirac, who had refined this wave equation. Dirac was the Lucasian Professor of mathematics at the University of Cambridge and had begun working on quantum mechanics as soon as Werner Heisenberg introduced it in the 1920s. His mathematics led to what he called a relativistic "theory of the electron" and then to his "theory of holes."

In his work, Dirac came up with a brand new equation, the Dirac equation, which had two solutions, and in so doing speculated that there must be a positively charged component within nature that balanced the negatively charged electron. Within two years of this suggestion,

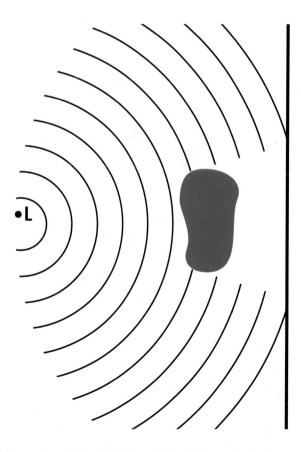

above *A figure from Schrödinger's Nobel prize speech. He was explaining that the wave-like property of light prevents clearly defined shadows even if the light comes from a point source.*

American physicist Carl David Anderson had found the positron, the antiparticle to the electron. This confirmed the theory, enabling Anderson to collect the 1936 Nobel Prize for Physics. The importance of Dirac's work lies in the fact that he introduced special relativity into quantum mechanics.

An expectation of beauty

One of the intriguing features about the way that scientists work is their basic assumption about how the universe operates. The assumption is that deep within its structure, it is beautiful—mathematically beautiful. They are also aware that it is fearfully difficult to grasp hold of, or even obtain a glimpse of, that beauty.

Commenting about his work in 1944, Schrödinger said: "I have no higher aim than to work out the beauty of science. I put beauty before science. *Nitimur in vetitum* [we strive for that which is forbidden]. We are always longing for our neighbor's wife and for the perfection we are least likely to achieve." This explanation is all the more descriptive of his mindset in that Schrödinger had numerous affairs with women throughout his life, some with the knowledge of his wife.

His sentiment for mathematics was echoed later by Dirac. "Schrödinger and I both had a very strong appreciation of mathematical beauty, and this dominated all our work. It was a sort of act of faith with us that any equations which describe the fundamental laws of nature must have great mathematical beauty in them," he said in 1977.

Of cats and boxes

If you want to carve your name in history, develop a theory that involves a cuddly animal. It probably wasn't Schrödinger's intention when he mentioned his "cat" thought experiment, but it is the feature of his work that has legendary status. The problem that Schrödinger was trying to explain was linked to radioactive atoms. There was an increasing realization that atoms can exist in more than one state at a time. It seems logically unachievable, but had to be the case if quantum theory was correct.

Schrödinger was aware that the idea of an atom simultaneously existing in both a decayed and a non-decayed state was difficult to comprehend, so he tried to illuminate the issue with a slightly strange thought experiment. He suggested that you considered the situation of a cat placed in a box. Before closing the box you put in a vial of deadly poison. The vial has a magical lid that contains a single radioactive atom. If the atom decays it releases a particle that triggers an electronic switch and opens the vial. The cat dies instantly. Because the atom can be both decayed and undecayed, the cat, says Schrödinger, is both dead and alive! "Schrödinger's cat" shows the mystery of the concept, rather than revealing any of the mechanism until we open the box to take a look, whereby it suddenly picks one or other state.

A 1996 experiment showed that a similar situation does indeed exist in nature. By supercooling an atom of beryllium and then prodding it with laser pulses the physicists managed to make it exist in two states simultaneously. The scientists described it as being like rolling a marble back and forth in a bowl and then finding that it suddenly appears as two marbles rolling in opposite directions, passing through each other and appearing simultaneously at each edge of the bowl. One thing is sure. Reality will never be as certain as it was.

Timeline

1887
Erwin Schrödinger is born on August 12 in Erdberg, Vienna, Austria, and was taught by a private tutor to the age of 10
1906
Enters the University of Vienna
1914
Publishes a paper developing some of Boltzmann's ideas. Schrödinger is called into the Austrian army and sent to the Italian border where he continues developing his theoretical ideas and submits another paper
1915
Transferred to Hungary, from where he sends another paper. Back on the Italian front line he receives a citation for outstanding service commanding a battery during a battle
1917
Sent to Vienna to teach a course on meteorology
1918–1920
Works on color theory
1920
Accepts assistantship in Jena, and marries Anny Bertel
1921
Takes position of professor of theoretical physics at Zurich
1925
Reads Louis de Broglie's doctoral thesis and then forms the first version of his theory of wave mechanics
1926
Publishes six papers relating to wave mechanics and the general theory of relativity
1927
Given post as professor of physics in Berlin, succeeding Max Planck (1858–1947) in the position
1933
Schrödinger shares Nobel Prize for Physics with Paul Dirac. In the same year he also leaves Germany and arrives in Oxford
1935
Schrödinger proposes his famous "cat" thought experiment
1936
Moves to Graz, Austria
1938
Nazis dismiss him from his post of theoretical physics in Berlin for "political unreliability" and he flees to Rome
1939
Moves to Dublin, Ireland, where he studies electromagnetic theory and relativity
1956
Returns to Vienna
1961
Dies on January 4 in Vienna

Werner Heisenberg

1901–1976

Acquaintances
— Wilhelm Wein (1864–1928)
— Arnold Sommerfeld (1868–1951)
— Albert Einstein (1879–1955)
— Max Born (1882–1970)
— Niels Bohr (1885–1962)
— Linus Wolfgang Pauli (1900–1958)
— Pascual Jordan (1902–1980)
— Paul Dirac (1902–1984)

It would be impossible, and indeed incorrect, to dissociate the physicists who were creating the foundations of quantum theory from the history of Europe in the first half of the 20th century. Much of the early work was centered in Germany, and the rise and fall of German fortunes caused emotional turmoil, interrupted scientific study, and meant that people were forced to move from place to place around the globe. There is no better example of this than Werner Heisenberg, a passionate patriot, who at times was revered by his country and at others decried. Against this background he revolutionized physics by being a key founder of quantum mechanics, and gained fame for his so-called "uncertainty principle."

Against a background of war

Werner Heisenberg was born in Würzburg, a city in the southern German state of Bavaria, on December 5, 1901. When Heisenberg was eight, his father became professor of middle and modern Greek philosophy at the University of Munich, so in 1910 the family moved to Munich. One year later he started school, but then World War I started. Through the war Heisenberg struggled to maintain his education, but was hampered because so many adults had been conscripted.

With the war at an end, Heisenberg appeared eager to help his country by assisting in the Bavarian agricultural service and took a leading role in Germany's youth movement. By 1920 a little normality was setting in and Heisenberg accepted a place at the University of Munich, where he studied physics with Arnold Sommerfeld, who immediately realized that Heisenberg was no ordinary student. In December 1921, with only 14 months of a university career under his belt, Heisenberg submitted his first paper for publication.

The fact that he found theoretical physics exciting and hated practical physics drove Heisenberg to develop mathematical approaches to problem-solving, but it came close to ending his career. In 1923 he presented his doctoral thesis and sat down in a room with four professors for the examination. Sommerfeld asked questions about the theoretical mathematics, and these were answered with ease. One of the examiners, Wilhelm Wein, was more concerned about practical physics and started to check that Heisenberg understood the details behind the experimental element of his work. Apparently he didn't. The result was a raging argument between Sommerfeld and Wein, the one wanting to pass him with flying colors, the other wanting to fail him. In the end a compromise was reached and Heisenberg was given a mediocre pass.

Doing business with giants

Although still in his early twenties Heisenberg found that he was meeting and arguing with the world's biggest names and refining their work. In 1924 he met Albert Einstein and a year later submitted his own paper on quantum mechanics. Inspired by his contact with Niels Bohr, Linus Wolfgang Pauli, and Max Born, he looked again at the current understanding of the atom, which speculated that electrons orbit a nucleus, much in the way that planets orbit the Sun. Working from observations of the emission and absorption of light by atoms he came up with a radically new concept. It was so new that he got Born to have a look at it before making it public. Born recognized that the equations belonged to a branch of mathematics that deals with arrays of numbers known as matrices. He was convinced — and quantum mechanics saw its first light of day.

From here to uncertainty

Most physicists disliked Heisenberg's use of matrix mathematics, because it was abstract and unfamiliar territory. Instead, they concentrated on Erwin Schrödinger's work. But in 1926 Schrödinger published proof that matrix and wave mechanics gave equivalent results — mathematically they were one and the same. However, Schrödinger maintained that his was the best way of looking at the subject. Needless to say, Heisenberg disagreed. Starting work as Bohr's assistant in Copenhagen he had plenty of opportunities to discuss the matter, as Schrödinger was a frequent visitor to the town.

At the same time, Paul Dirac was working on the issue with colleague Pascual Jordan. They created unified equations known as "transformation theory." These were all very nice on paper, but Heisenberg wanted to know what they meant to a real live atom.

The more he studied the mathematics, the more he perceived a problem. If you pinned down the position of an electron, you couldn't say anything about its momentum.

above *A group portrait of eight physicists and Nobel Laureates. Werner Heisenberg is on the far right.*

above *Werner Heisenberg teaching in 1936. He was a founder of quantum physics, being awarded a Nobel Prize in 1932 for his matrix mechanics version of the theory of quantum physics.*

Conversely, detect an electron's momentum and you can't measure its position. This, he said, wasn't a fault of the equations, but reflected a fundamental property of quantum mechanics. Heisenberg presented this to Pauli in a 14-page letter in 1927 and then having discussed and refined the ideas with Bohr, he subsequently presented it to the world. Heisenberg's uncertainty principle had arrived.

Friction and fusion

Having created a stir, people were keen to meet Heisenberg, and he spent the best part of 1929 traveling to the United States, Japan, China, and India, but trouble was brewing at home. In 1933 he belatedly received the 1932 Nobel Prize for Physics and witnessed Adolf Hitler come to power in Germany. Four years later Heisenberg and theoretical physics were attacked in a Nazi party newspaper. Life was going to get tough.

Even in troubled times normal life continues, and in 1937 Heisenberg married Elisabeth Schumacher in Berlin. In the same year an S.S. newspaper launched another attack on his work. The following year brought only good news as he witnessed the birth of twins, the first of seven children, toured England, and then heard that Himmler had exonerated him of S.S. charges. The reason for this is clear. The S.S. had recognized the potential of nuclear fission.

With hostilities looming, Heisenberg bought a home in Urfeld, Bavaria, as a retreat for his family, and after the outbreak of war he joined the fission research project in Berlin, keen to develop the power for his homeland. It was a politically sensitive project, which on occasion found favor with the authorities and at other times almost had its funding cut. The war, however, came to an end before the project was complete and in May 1945 Heisenberg was arrested at his home in Urfeld by U.S. forces a couple of days before the German surrender. Politics had stimulated his work — now it stopped it.

Heisenberg was held with other German scientists at Farm Hall, England, and would have been forgiven for believing that his research was over. But fortunes change and in January 1946 he was released and allowed to return to Germany, where he settled in Göttingen. Keen to rebuild German science, he eagerly took the posts of director of the Kaiser Wilhelm Institute for Physics and the founding president of the German Research Council.

For the last 25 years of his life Heisenberg generated new ideas, while taking an active political role in international science. He headed the German delegation to the European Council for Nuclear Research when it was establishing CERN — the European Organization for Nuclear Research — and in 1957 he joined 17 other West German scientists in issuing a declaration opposing the Chancellor of the West German Republic Konrad Adenauer's acceptance of NATO tactical nuclear weapons.

In 1970 he resigned as director of the Max Planck Institute and in 1975 he gave up the presidency of the Alexander von Humboldt Foundation. On February 1, 1976, he died of cancer.

The Uncertainty Principle states

The more precisely the position is determined, the less precisely the momentum is known in this instant, and vice versa.

One of the primary goals of physics is to draw together the understanding of different events and show how they relate to each other. Isaac Newton did it when he created a unified theory for why apples fall to the ground and planets orbit the sun. James Clerk Maxwell did it when he unified the theories of electricity and magnetism, and Albert Einstein when he drew together time, space, and gravity. Einstein died, however, still searching for a way of unifying general relativity with electromagnetism. Tackling this challenge became a central aspect of Abdus Salam's work, as he found ways to unify the weak nuclear force and the electromagnetic force.

Abdus Salam

1926–1996

Acquaintances
— Paul Dirac (1902–1984)
— Sheldon Lee Glashow (1932–)
— Steven Weinberg (1933–)

The Standard is more than meets the eye

The more physicists study the fundamental nature of matter, the more complex and confusing it tends to become. In Newton's day events were puzzling, but everyone concentrated on physical units that you could see and often literally handle. All that has changed.

Theoretical physics has become a mathematical pursuit that makes predictions about the way things might be and balances these predictions with the data pouring from multi-million dollar high-energy particle accelerators. The results at times seem fanciful, but the best understanding at the dawn of the 21st century is called the Standard Model, and owes much to the work of Abdus Salam, who along with Steven Weinberg and Sheldon Lee Glashow, was awarded the 1979 Nobel Prize for Physics. The Model talks in terms of quarks, electrons, neutrinos, gluons, and Higgs particles.

The discovery of radioactivity of certain heavy elements, and the ensuing development of the physics of the atomic nucleus, led physicists to introduce two new interactions — the strong and weak nuclear forces. Unlike gravitation and electromagnetism these forces only operate at very short distances, distances of the order of the diameter of the nucleus or less. Weak interactions change neutrons into protons and vice versa in radioactive processes within stars. They are responsible for the so-called radioactive beta-decay. Strong forces hold quarks together inside protons and neutrons, as well as holding protons and neutrons together inside atomic nuclei.

The Standard Model of particle physics has unified electromagnetism with the weak interactions, and provides an explanation of the strong interactions. There are still huge holes in the theories, for example there is no way yet of creating a unified understanding of strong and weak interactions.

Weak makes the sun shine

The electroweak theory that Salam introduced now sits at the core of the Standard Model of high energy physics. Like much theoretical physics it took a few years from the first time it was mentioned until it was proved to be correct in some form of physical experiment. In 1973, physicists at the superprotosynchrotron facility at CERN in Geneva, discovered W and Z particles, both predicted and indeed required by the theory.

Like Planck's constant, the strength of the weak interaction appears to be fundamental to the way the universe operates, and as such is critical to our own existence. For example, life on earth is powered by energy in sunlight. This is produced when hydrogen fuses, or burns, into helium in a chain of nuclear reactions occurring in the Sun's interior. In the first step of the chain reaction, weak forces enable hydrogen to transform into heavy hydrogen (deuterium). Without this force there would be no solar energy. In addition, if the force had been any stronger, the life span of the Sun would have been too short to give

time for life to evolve. Any smaller and the energy from the Sun would have been insufficient.

An interest in peace

Although his ability to remodel science had given Salam many rewards and privileges, he never forgot his roots, and devoted much of his life to the goal of international peace and cooperation. He had particular concern for the increasing gap between developed and developing nations, believing that this disparity will never be reduced until developing countries establish their own scientific and technological industries.

In establishing the ICTP in Trieste, he sought to enable students from disadvantaged backgrounds to experience life in the most prestigious research environments, and to take that learning back to their own countries.

His life as a devout Muslim was also an integral part of his work. He once wrote: "The Holy Quran enjoins us to reflect on the varieties of Allah's created laws of nature; however, that our generation has been privileged to glimpse a part of His design is a bounty and a grace for which I render thanks with a humble heart."

Timeline

1926
Born in Jhang, a small town in what is now Pakistan
1940
Gains highest ever marks in his matriculation examination at the University of Punjab, age 14
1946
Receives MA and is awarded a scholarship to St. John's College, Cambridge
1951
Publishes his Ph.D. thesis, which contains fundamental work in quantum electrodynamics, and returns to Pakistan to teach mathematics at the Government College in Lahore
1952
Becomes head of mathematics at Punjab University
1954
Takes lectureship in Cambridge
1957
Becomes professor of theoretical physics, Imperial College, London
1961–1974
Acts as Chief Scientific Advisor to the President of Pakistan
1964
Establishes the International Center for Theoretical Physics (ICTP) at Trieste, Italy, and uses his position as director to create "Associateships" that enable students from developing countries to spend three months each year working in a Western research establishment
1996
Dies in Oxford, England

Democritus

c.460–c.370 B.C.

Acquaintances
— Anaxagoras (c.499–c.428 B.C.)
— Leucippus (c.480–c.420 B.C.)

The first philosophers were obsessed with trying to find explanations for their existence. Folklore and mythology told stories of gods and spiritual forces, but many found that these were deeply unsatisfactory. Democritus became the focus for a small band of people who developed a materialistic view of life, the universe, and everything. All that you can see, indeed all that you are, they claimed, is a product of packing together a myriad of minuscule atoms. There is only one type of atom and any variation that you see is due to differences in the ways that these atoms are grouped.

To be or not to be

There are some people who leave an imprint on history while leaving very little evidence of their own life. Democritus is one of these. Born some time around 460 B.C. in Abdera, Thrace, in northern Greece, he appears to have traveled widely, spending time in India, Egypt, and Persia, all great centers of academic thought. Some stories suggest that he financed the trip with money left to him when his father died.

Some reports of his life have him boasting that he had "seen more climes and countries and listened to the greatest number of learned men" than anyone else. He appears therefore to have been disappointed that no one was watching, because, according to the second-century A.D. writer Diogenes Laertius, on visiting Athens he found that no one knew him and the ageing philosopher Anaxagoras apparently refused to see him.

So why has someone, so unknown during his day, become considered to have provided a milestone in philosophy? The answer is that Aristotle disagreed with his theories, and thus wrote about them, and conversely the Greek philosopher Epicurus liked the ideas and so adopted them within his work.

At the center of Democritus' ideas was the concept that everything is made of atoms, and that the world ran purely on mechanical properties. There is only one type of atom and any variation that we see as we compare, say, granite and silk, or water and ice, derives from differences in the ways that the atoms are stacked. The atom is an indivisible, solid object. It is absolutely small, but is completely full and cannot be compressed. This ran counter to mainstream thinking, exemplified by the philosophies of people like Anaxagoras, that matter was infinitely divisible. As is more often than not the case, history would show that they both were partly correct.

Initially there was a problem with atomism. If atoms were the only thing in the universe, then how come there is any space between objects? After all, with nothing but atoms the entire universe would be one massive solid lump, like an infinite box of sugar cubes. To solve this, Democritus developed the second component of the atomist concept, namely the void. Balancing the idea that the atom is the only unit of being, the void is the single kind of not-being. The void is the empty extension that reaches out from atoms.

Atomistic mathematics

This atom theory had no origins in scientific experiment, but came purely from following a pattern of thought. This doesn't mean that Democritus ignored practical issues.

One of the dilemmas that he presented was how to deal with a cone. He had correctly worked out that the volume of a cone was one third of the volume of a cylinder that had the same height and base diameter. He was, however, troubled by the make-up of a cone. If, he asked, you cut through a cone parallel to its base, would the circle formed on the top section be the same size as that on the bottom of the new smaller cone?

If it were, then the cone would in fact be a cylinder and clearly that was not true. If they were not equal, then the surface of a cone must consist of a series of steps or indentations. This fitted better with his atomic theory, but he had to concede that he had not seen steps on the surface of a cone.

An atomic soul

Life, for Democritus, was nothing but a collection of atoms that existed and operated together when grouped into a body. The soul was nothing more than a special group of atoms that were moving very fast and operated within living beings to help them sense the environment and act in appropriate ways.

Disperse the atoms and you have nothing left. The soul is therefore destroyed by death, just as water is lost from a vessel if it is smashed on the floor. There is no place for immortality within Democritus' atomistic view of life. This marked the arrival of materialism, the idea that we are no more than the material of which we are made.

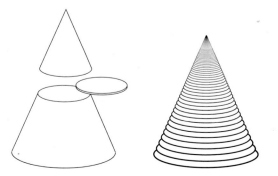

left *Democritus was convinced that everything was made of finite atoms. A pure cone was therefore a stack of discs that get smaller, until a single atom forms the pointed top.*

A sick child from an exceedingly wealthy Irish family, Robert Boyle had a truncated formal education but an enquiring mind. His passion was alchemy, but paradoxically his work set the foundations for modern chemistry. Working with Robert Hooke he concluded that the volume, pressure, and temperature of gases were intrinsically linked, in a manner that has become known as Boyle's law. His investigation of matter led him to challenge Aristotelian principles and promote the atomic concepts, with the additional idea that atoms could combine to form molecules. He was rich enough not to need a job, but accepted the post of director of the East India Company so that he could propagate Christianity around the world.

1627–1691   **Acquaintances**
—Robert Hooke (1635–1703)

Robert Boyle

Atoms and molecules

Boyle was born in Lismore Castle, Munster, Ireland, on January 25, 1627, and died in London on December 30, 1691. In between, he had given us a mechanical understanding of the behavior of gases, and pointed the way toward an understanding of matter that moved scientists away from the Aristotelian view that everything was made of four elements — earth, air, fire, and water.

In 1661 he published his book entitled *The Sceptical Chymist*, in which he presents the idea that elements are "primitive and simple, or perfectly unmingled bodies." These elements, he believed, were capable of combining to make compounds. The elements were still present because those compounds could be split back into their elements.

The book takes the form of a dialogue between four characters. Boyle represents himself in the form of Carneades, a person who does not fit into any of the existing camps, as he disagrees with alchemists and sees chemists as lazy hobbyists. Another character, Themistius, argues for Aristotle's four elements, while Philoponus takes the place of the alchemist, and Eleutherius stands in as an interested bystander.

In the conclusion he attacks the chemists for playing games rather than doing serious science. "I think I may presume that what I have hitherto Discursed will induce you to think, that Chymists have been much more happy finding Experiments than the Causes of them; or in assigning the Principles by which they may best be explain'd."

On the same page he pushes the point further: "me thinks the Chymists, in the searches after truth, are not unlike the Navigators of Solomon's Tarshish Fleet, who brought home from their long and tedious Voyages not only Gold, and Silver, and Ivory, but Apes and Peacocks too; For so the Writings of several (for I say not, all) of your Hermetick Philosophers present us, together with divers Substantial and noble Experiments, Theories, which either like Peacocks' feathers made a great show, but are neither solid nor useful; or else like Apes, if they have some appearance of being rational, are blemish'd with some absurdity or other, that when they are Attentively consider'd, makes them appear Ridiculous."

The critical message from the book was that matter consisted of atoms and clusters of atoms. These atoms moved about, and every phenomenon was the result of collisions of the particles. He appealed for high-quality experiments and called for chemistry to free itself from its subservience to either medicine or alchemy.

Pressure, volume, and temperature

Just before Boyle published *The Sceptical Chymist*, he had announced the conclusion of some work that he had done together with his assistant Robert Hooke. Hooke had developed an air pump, and that gave Boyle the opportunity to look at the mechanical properties of gases.

Together they put a burning candle in a jar and then pumped the air out. The candle died. Glowing coal ceased to give off light, but would start glowing again if the air was let in while the coal was still hot. Clearly combustion required some physical property of the air. In addition, they placed a bell in the jar and again removed the air. Now they could no longer hear it ringing.

Many of Boyle's results puzzled him, but one thing he did clarify, and that was the relationship between the temperature, pressure, and volume of a fixed amount of gas. If you squeezed the gas and decreased its volume, the temperature rose. If you increased the temperature and held the volume constant then the pressure increased. Hence Boyle's law states that for a fixed mass of gas, pressure and volume are inversely proportional.

Intermingled beliefs

A remarkable feature about *The Sceptical Chymist* was that Boyle published it at all, and it is fair to assume that the publication would have made enemies. He was an alchemist and alchemists worked in secrecy. Ideas were passed only to members of the "club." The book would have been particularly shocking as it not only discussed the philosophy of the science, but also presented detailed descriptions of the experiments.

The English antiquarian John Aubrey alluded to the way that Boyle would use his wealth to obtain knowledge. "He is charitable to ingenious men that are in want, and foreign chemists have had large proof of his bounty, for he will not spare for cost to get any secret."

In fact, Boyle held a complex mixture of views. In addition to his science and his alchemy he was a passionate believer in the need to teach people about Christianity, and by bequest founded the "Boyle lectures" in defense of Christianity. He took the job as director of the East India Company primarily, it would appear, so that he could influence its involvement in propagating Christianity in the East.

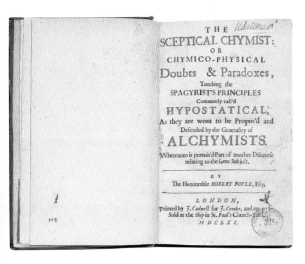

above *Publication of* **The Sceptical Chymist** *marked the start of a new era in the thought patterns of scientists who were grappling to make sense of the physical world.*

Antoine Lavoisier picked up where Boyle left off. He radically reformed the concept of chemistry, and gave it a new language and fresh goals. He killed off the Aristotelian concept that elements could transmute into each other and provided a firm foundation for Dalton's atomic theory a few decades later. He proved that water cannot be converted into earth, showed that metals burn by absorbing "vital principle" from air, demonstrated that water is made of hydrogen and oxygen, and attacked the theory of phlogiston. Sadly, he was executed during the French Revolution.

Antoine Lavoisier

1743–1794

Acquaintances
— Jean Etienne Guettard (1715–1786)
— Joseph Priestley (1733–1804)
— Louis-Bernard Guyton de Morveau (1737–1816)
— Claude-Louis Berthollet (1748–1822)
— Pierre Simon de Laplace (1749–1827)
— Antoine Fourcroy (1755–1809)

The death of phlogiston

The son of a prosperous lawyer, Antoine Lavoisier studied law, but his essay on a method of providing street lights was so exceptional that aged only 25 he was elected to the French Royal Academy of Sciences. In the same year he bought shares in a tax-collecting company, a move that helped finance his work, but ultimately cost him his life.

Aged 28 he met and married the 14-year-old daughter of another shareholder of the tax company, and for the rest of his life Marie-Anne became an invaluable assistant and companion in his scientific work.

While still studying law, Lavoisier had been captivated by a popular chemistry course taught by François Rouelle in the lecture rooms at the Jardin de Roi. This fired his imagination and set the course of his intellectual pursuit.

His first task was to debunk the myth that water could turn into earth. The idea had developed from a number of observations, but the most controlled was that if chemists repeatedly distilled water, they found a residue of material building up. By carefully weighing his apparatus, Lavoisier proved that this residue had dissolved out of the glass that made up the distilling equipment. Nothing strange had happened to the water.

His next assault was on the theory of phlogiston. German chemist Georg Stahl had produced a theory of matter, stating that all substances were composed of water and three varieties of earth, one of which was a combustible material called phlogiston. Metals, he said, burned in air not because they absorbed anything from the atmosphere, but because they lost phlogiston. The idea appeared to be confirmed by the observation that metals could be recovered from the residue, the "calces," by combining them with a phlogiston-rich material like charcoal.

Intial observations also showed that when you burned a metal by making it very hot, the calces were lighter than the original sample. Clearly the process had driven off the phlogiston. The first serious nail in the coffin came when a French part-time chemist, Louis-Bernard Guyton de Morveau, reported that if you measured things carefully the calces were actually heavier than the original metal. To Lavoisier this meant one thing. When metal burnt it must be combining with something from the air, not releasing something to it. And he proved it by using a huge lens to concentrate sunlight onto a mixture of an oxide of lead — a calx — and charcoal. As it heated, a large volume of "air" was given off and the lead was left behind.

Unfortunately, Lavoisier was unaware of the work that Joseph Priestley was doing in England. Priestley had discovered that there were many types of air, and atmospheric air consisted of a mixture of different gases. On visiting Paris in August 1774, Priestley showed Lavoisier that if you heated some calces of mercury in an enclosed vessel it returned to mercury without the need for charcoal. Clearly this reaction needed no source of phlogiston. Furthermore, the gas that collected in the vessel could support combustion better than ordinary air.

It took until 1778 before Lavoisier worked out what was going on. We now know that he had discovered oxygen.

The birth of chemistry

Earth may not have come from water, but you could argue that modern chemistry did. Priestley had noticed that if you sent an electric spark through a mixture of air and "inflammable air" (hydrogen) there was first of all an explosion, and then moisture — he called it dew — was left behind. Priestley was sure that this dew was pure water. Assisted by Laplace, Lavoisier then found that burning inflammable air and oxygen in a closed vessel also produced water. On the other hand, by placing iron filings in water he discovered that they rusted and released the inflammable air.

The key importance of this was that Lavoisier became convinced that water was a compound made of two different elements. The elements could be pulled apart to give separate gases, or combined to give water.

Working together with Guyton, Count Claude-Louis Berthollet, and Antoine de Fourcroy he developed a new language to describe their work — a language that did away with any concept of phlogiston or alchemy. He defined a chemical element, saying that it was any substance that could not be analyzed further. With this definition he identified 33 elements, though some have since proved to be compounds.

Timeline

1743
Born in Paris on August 26

1770
Proves that water cannot be transmuted into earth

1771
Marries the 14-year-old Marie-Anne Pierrette Paulze

1772–1775
Carries out experiments that show that metals burn by absorbing vital principle from air

1776
Moves to live in the Royal Arsenal and takes charge of the production of gunpowder

1776
Erroneously decides that all acids contain oxygen

1785
Attacks the theory of phlogiston

1783
Demonstrates that water is a compound of hydrogen and oxygen

1787
Defines chemical elements and publishes *Method of Chemical Nomenclature*

1789
Along with three other French chemists, he publishes *Elementary Treatise of Chemistry*

1793
Lavoisier is arrested on November 24, along with the other owners of the tax company

1794
Lavoisier is executed by guillotine on May 5, in Paris

Developing an understanding that material is made of many different types of atoms was a critical insight and one that gave birth to the science of chemistry. It took the patience of an obsessive bachelor and devout Quaker. John Dalton studied the properties of gases and began to provide a framework that would enable scientists to make sense of chemical reactions and make accurate predictions of the outcome of mixing different types of atoms. His system of notation enabled a new type of note-taking, but also enshrined the concept that matter was built of atoms, even though they had not yet been actually discovered.

John Dalton

1766–1844

Acquaintances
—John Gough (1757–1825)

An elemental interest

When John Dalton started investigating chemicals and chemical reactions he entered the science of a raging debate. It was a debate that hadn't been resolved since the days of Democritus and Aristotle. Was everything built of finite building blocks — atoms — or was it infinitely divisible? Tagged onto this was an associated argument as to whether you could define rules for the ways chemicals behave, or whether you could only draw generalizations about nature. The prevailing view tended to be that atoms were simply a convenient way of describing things, but in reality material was infinitely divisible, and that you would never pin nature down with rigid rules of behavior. Dalton was about to turn both of these on their heads.

The behavior of gases was Dalton's first point of interest, and he soon developed what became known as Dalton's law. This states that the pressure of a mixture of gases is equal to the sum of the pressures of the gases in the mixture. He realized that each gas acted independently of the other.

On heating gases he discovered that they expanded, and he showed that gas could dissolve in water or diffuse through solid objects. In the second chapter of his book *A New System of Chemical Philosophy*, he describes how his work on gases led him to conclude that everything was made of atoms and points out that water can exist as an elastic fluid, liquid, or a solid.

In 1803, while studying the way that nitrogen and oxygen combine, he came to the conclusion that the two gases always joined together in one of two different whole number ratios. He had the right idea, although he got the exact numbers wrong.

A novel notation

Also in 1803, Dalton developed a system of notation that enabled him to write down what was occurring when elements combined. Oxygen appeared as an empty circle, while hydrogen was drawn as a circle with a dot in the middle. Water, which is made of two atoms of hydrogen and one of oxygen, is therefore written as two circles with dots and one open circle. It is the forerunner of the modern H_2O.

above *This booklet of threads dates back to the early 1800s and was used by Dalton to test his own color-blindness.*

One of Dalton's hobbies was lawn bowling and he believed that atoms were very similar to miniature bowls. Drawing them as circles made sense of the concept, and he was wary of any attempt to represent them in letters. The debate had real meaning. Some people argued that the atoms should have letters because this could still be interpreted as a generalized concept. Dalton liked his graphic representation because it emphasized his theory that the atoms were real physical units.

Honoring a humble man

Throughout his life Dalton chose to live simply. It was part of his Quaker faith. On his death, however, he could have no say, and was laid in state in Manchester town hall where over 40,000 people paid their last respects. Contemporary reports say that the funeral procession consisted of one hundred coaches and was more than a mile long. It was a lavish event that embarrassed his Quaker friends.

His final experiment was carried out after his death. In the middle of his life he and his brother discovered they had color-blindness. In his will he asked for his eyes to be examined to see if there was a physical cause. No cause was found and physicians assumed that it was due to some deficiency in his sensory powers.

Timeline

1766
Born on September 6
1787
Dalton starts to supplement his income by giving public lectures
1792
Both Dalton brothers discover that they are color-blind
1793
Moved to Manchester as tutor at New College, an establishment founded by Presbyterians. At that point, education at Oxford or Cambridge was only available to members of the Church of England. He publishes his first book *Meteorological Observations and Essays*
1802
Dalton publishes a book in which he states his law of partial pressures of gases
1803
Dalton announces his law of multiple proportions and publishes his list of atomic weights and symbols
1808
Publishes *A New System of Chemical Philosophy — Part I*
1808
Publishes *Part II*
1817
Becomes president of the Philosophical Society, a post he holds until his death
1844
Dies of a stroke on July 27, having, as usual, entered the day's weather conditions in his daily journal

Robert Boyle and Antoine Lavoisier had seen glimpses of an idea, but Edward Frankland started to make sense of it. Atoms when they came together to make compounds did so in regular ratios. This was the theory of valence, with the valence number being the number of chemical bonds that any given atom can make with other atoms when forming a compound. The concept forms the foundation of modern structural chemistry. He also introduced the term "bond" to describe the way that atoms link. This was followed by Friedrich Kerkulé's realization that carbon can form chains and rings, thus giving rise to organic chemistry.

Edward Frankland

1825–1899

Acquaintances
— Robert Wilhelm Bunsen (1811–1899)
— Friedrich August Kerkulé (1829–1896)
— Alexander Crum Brown (1838–1922)

A theory of valence

Quite often in science, a person starts work on one concept only to reveal something more interesting in another field. Occasionally it is something fundamentally important. This was to be the case with the English chemist Edward Frankland.

While working with Robert Bunsen in Germany he became fascinated by a class of chemicals that bound metal atoms with other compounds, now called organometals. The particular set of these that he was looking at were zinc dialkyls.

As well as an object of study, it appears that he used these compounds for entertainment. Frankland describes that, when he added water to these compounds, "a greenish-blue flame several feet long shot out of the tube, causing great excitement amongst those present and diffusing an abominable odor through the laboratory."

When he was being serious, however, he spotted a pattern. As is often the case it was a pattern that had been seen before. A few years earlier, Alexander Crum Brown had seen it as well and reported it in his M.D. thesis at the University of Edinburgh, entitled "On the Theory of Chemical Combination." In this thesis he wrote: "It does not seem to me improbable that we may be able to form a mathematical theory of chemistry, applicable to all cases of composition and recomposition."

What both Crum Brown and Frankland had seen was that when elements combine, they do so in whole number ratios. Lavoisier had been moving toward the idea when he split and recombined water and found constant proportions of hydrogen and oxygen were always involved. Frankland took it further and developed what he initially called atomicity, but we now know as valence. In his first report of it he said, "The combining power of the attracting element ... is always satisfied by the same number of atoms..."

The idea is that every atom has a fixed number of bonds that it can form, and that to be stable, all of these must be employed. With the benefit of more information we know that, for example, hydrogen has one and oxygen has two. This gives few options in the ways that these two atoms can combine. If a hydrogen atom bonds to another hydrogen atom, then the bonds on each atom will be fully used in forming H_2, otherwise known as a molecule of hydrogen. The same can occur between two atoms of oxygen.

Alternatively, the two bonds on oxygen could be occupied by the bonds on two hydrogen atoms, forming H_2O — or water. Frankland realized that only molecules in which atoms had all of their bonds occupied were stable, or for that matter "possible."

Frankland also introduced the notation that is now so commonly used to express the structure of molecules that it is easy to forget that it had to be invented. In this notation, water, or H_2O, is drawn as H-O-H.

The concept of valence was picked up and developed a few years later by Friedrich August Kerkulé, who claimed to have had a vivid daydream while traveling on the upper deck of a London bus. Kerkulé decided that the valence of carbon must be four. Not only this, but he suggested that this would allow carbon to form into chains of atoms, thus creating huge molecules. It was a radical idea, and more importantly it was right.

In 1865 Kerkulé went on to propose that carbon could not only form chains, but could also link into closed six-atom rings. In the simplest such molecule, three of each carbon's bonds are used to keep the ring together and the remaining bond on each carbon binds to a hydrogen atom. The resulting molecule contains six atoms of carbon and six hydrogen atoms and is known as benzene. Understanding benzene's structure launched a massive explosion in the chemical engineering of materials, with uses ranging from dyes to drugs.

Thus Frankland and Kerkulé together enabled chemistry to become a tool to create new molecules, and not simply a method of observing what already exists.

Timeline

1825
Born on January 18 in Churchtown, Lancashire, the illegitimate son of a distinguished lawyer

1840
Becomes an apprentice in a chemist's shop

1847
Becomes the teacher of chemistry at Queenwood College, Hampshire, but soon moves to Marburg in Germany to work for three months with Robert Bunsen, during which time he discovered the science of organometallic compounds

1848
Returns to London

1849
Completes his Ph.D.

1851
Becomes professor of chemistry at Owen's College, Manchester, now Manchester University

1852
Pointed to the "general symmetry" of the formulae of a number of chemical compounds

1857
Moves to St. Bartholomew's Hospital, London

1865
The government asks Frankland to analyze metropolitan drinking water

1868
Becomes professor of chemistry at the Royal College of Chemistry

1880
Publishes *Water Analysis for Sanitary Purposes*

1885
Retires, but carries on working on the chemistry of storage batteries

1897
Knighted

1899
Dies on August 9 after a brief illness that started while on a salmon-fishing trip to Norway

It is likely that Aristotle would have loved the development in chemistry that is commonly attributed to Dmitri Mendeleev—the periodic table. This is arguably the single most significant exercise in assigning items to groups that has ever occurred. In a moment of inspiration, Mendeleev listed elements according to their known weights, and recognized repeating patterns in their behavior. He had the confidence in his identified pattern to leave gaps in the list, anticipating that they would be filled by future discoveries. He was right: the "missing" elements have all been found.

Dmitri Mendeleev

1834–1907

Acquaintances
- Johann Dobereiner (1780–1849)
- Alexandre Beguyer de Chancourtois (1820–1886)
- Julius Lothar Meyer (1830–1895)
- John Newlands (1837–1898)
- Lord Raleigh (1842–1919)
- William Ramsey (1852–1916)
- Henry Mosely (1887–1915)

Playing elemental poker

Born in Serbia, Dmitri Mendeleev was the youngest in his family and grew up with his 13 brothers and sisters. When his father, a teacher, went blind, his mother started a glass factory to generate an income. Tragically, just as Mendeleev turned 14 his father died and the factory burnt down, forcing the family to move to St. Petersburg. There, Mendeleev managed to complete his education.

He was a bright student and rose through the academic ranks to become professor of chemistry at St. Petersburg, where he set about his work looking at similarities in the behaviors of different elements.

In doing this he would have been aware of the law of triads developed by German chemist Johann Dobereiner. Dobereiner had recognized mathematical patterns in elements that had similar properties. He found that if you added the atomic weight of calcium (40) to that of barium (137) and divided this value in two, you were left with a value very close to the weight of strontium (88). When Dobereiner found that the same pattern occurred for lithium, sodium, and potassium, and for chlorine, bromine, and iodine he was convinced that this was more than coincidence. He named his idea the Law of Triads but failed to get much further. As he worked, Mendeleev might have been aware of the work of the French scientist Alexandre Beguyer de Chancourtois, who, in 1862 developed a way of representing the elements by wrapping a helical list around a cylinder. Working at the School of Mines in Paris, he published his ideas without including a diagram. Sadly few people understood what he was trying to say, and the work was largely ignored.

According to Mendeleev's notes, his table came as a moment of genius-driven inspiration, assisted by a pack of cards. In a single day, February 17, 1869, Mendeleev sat with a pack of 63 cards. On each card was written the name of an element, its atomic weight and any known physical and chemical properties. This pack contained all of the elements known at that time.

John Dalton had proposed in 1805 that each element had a characteristic atomic weight. By sorting the cards around, Mendeleev found a way of placing the cards in a grid where the atomic weight increased as you went down a column, and the elements in any row shared similar properties. The first column started with lithium, followed by beryllium, boron, carbon, nitrogen, oxygen, and fluorine. In a modern periodic table, this set of elements appears as the first row, as the table has been turned sideways.

For this to work, he had to juggle the order of a few of the elements, but was content to do that because he rightly assumed that their weights must have been incorrectly measured. At the same time he had to leave three prominent gaps for elements that he assumed must exist, and, moreover, drew up specifications of how he would expect these elements to behave once they were discovered.

By 1886 all of these three elements had been found and named as gallium, germanium, and scandium. In addition, chemists discovered that their properties matched Mendeleev's predictions. Coming up with a way of explaining what you can already see is clever, but using that explanation to make predictions that turn out to be correct adds huge credit to the whole idea.

It would take another half century before the rationale underlying Mendeleev's observational work would gain a foundation in the concept of orbits, introduced by Niels Bohr.

No prizes this time

It appears that Mendeleev narrowly missed out on being awarded a Nobel Prize for this piece of work. Some reports say that in 1906 he came within one vote of receiving the honor. His contribution was, however, recognized when element number 101 was found, and given the name mendelevium. He found himself frequently fighting internal university political battles and in the end resigned from the University of St Petersburg in 1890, but after a few years he was appointed director of the Bureau of Weights and Measures in St. Petersburg, a post he retained until his death from pneumonia on January 20, 1907.

below *The modern periodic table has its foundations in Mendeleev's recognition of patterns in the differing atomic weights of elements.*

1 H																	2 He
3 Li	4 Be											5 B	6 C	7 N	8 O	9 F	10 Ne
11 Na	12 Mg											13 Al	14 Si	15 P	16 S	17 Cl	18 Ar
19 K	20 Ca	21 Sc	22 Ti	23 V	24 Cr	25 Mn	26 Fe	27 Co	28 Ni	29 Cu	30 Zn	31 Ga	32 Ge	33 As	34 Se	35 Br	36 Kr
37 Rb	38 Sr	39 Y	40 Zr	41 Nb	42 Mo	43 Tc	44 Ru	45 Rh	46 Pd	47 Ag	48 Cd	49 In	50 Sn	51 Sb	52 Te	53 I	54 Xe
55 Cs	56 Ba	57 La	72 Hf	73 Ta	74 W	75 Re	76 Os	77 Ir	78 Pt	79 Au	80 Hg	81 Tl	82 Pb	83 Bi	84 Po	85 At	86 Rn
87 Fr	88 Ra	89 Ac	104 Rf	105 Db	106 Sg	107 Bh	108 Hs	109 Mt									

58 Ce	59 Pr	60 Nd	61 Pm	62 Sm	63 Eu	64 Gd	65 Tb	66 Dy	67 Ho	68 Er	69 Tm	70 Yb	71 Lu
90 Th	91 Pa	92 U	93 Np	94 Pu	95 Am	96 Cm	97 Bk	98 Cf	99 Es	100 Fm	101 Md	102 No	103 Lr

Just as the majority of scientists were coming to terms with the notion that atoms existed and were indivisible, along came Marie Curie and her husband, Pierre. Together they produced more evidence showing that atoms existed, and that they were not immutable. They could release particles and in so doing change. Her work was remarkable, in that she not only discovered radiation, but also found ways of putting this novel source of power to use. She was awarded two Nobel Prizes, which is a remarkable feat, as at that time women seldom gained recognition for their work.

Marie Curie

1867–1934

Acquaintances
—Gabriel Lippmann (1845–1921)
—Antoine Henri Becquerel (1852–1908)
—Henri Poincaré (1854–1912)
—Albert Einstein (1879–1955)

A ray of inspiration

When you need to introduce a new idea, you need a rebel who is prepared to fight. In Marie Curie we find the qualities. Born in Poland, she followed her sister to Paris to study at the Sorbonne. As one of 210 women among nine thousand male students, she was already marked as unusual, and when she did better than all of them in examinations she was recognized as being bright.

Rather than living with her sister in "Little Poland" she chose to break with form and move into an apartment in the vicinity of the Sorbonne. Aged only 23, this was a shocking display of independence for a woman.

Being bright is one thing. Revolutionizing science is quite another. Shortly after graduating she married renowned physicist Pierre Curie. It was a relationship that was initially based on a mutual interest in science, but which prospered. Curie wrote once: "I have the best husband one could dream of. I could never have imagined finding one like him …"

Under Pierre's inspiration, Marie Curie started investigating some curious findings that physicist Henri Becquerel had made, before moving on to study other things. Becquerel had found that uranium gave out strange rays, and called them Becquerel rays. In studying his work, Curie found that the natural ore pitchblende gave out more of the rays than did uranium. At first she didn't believe her results, but then went on to discover that pitchblende contained two radioactive elements in addition to uranium. Sack by sack, the Curies stubbornly hauled tons of pitchblende into their laboratory, and slowly extracted the new elements, radium and polonium.

Changing the unchangeable

The existence of these elements was one thing. It was what they did that was so curious. At the end of the 19th century the nature of matter was still under debate, and Curie was a member of the group of scientists who believed matter was composed of atoms.

Classical atom theory then said that these atoms were the smallest possible unit and that they could not be divided into other bits and pieces. Curie showed that this was not the case and that they could change. First of all, Curie found that the rays leaving the radioactive elements were particles released by the atoms, which showed that the atoms were capable of dividing into smaller units. And secondly, when the particles left the atom, the atom changed. It became a new element. The alchemists would have loved it.

Winning the Nobel Prize changed the Curies. It was only the third year that the prize had been awarded, and it gave them fame and recognition. It also gave them access to funds, because Pierre was admitted to the Académie des Sciences, an important source of money for research in France.

One thing, however, didn't change. Marie was not offered a place at the Académie. Discrimination against women was still very much alive. Indeed, the initial recommendation for a Nobel Prize didn't include her name. It was added at Pierre's insistence.

The medical wonder and nightmare

Marie Curie was all too aware of the usefulness of radiation, and during World War I worked hard to mobilize X-ray units to the front line to assist in the treating of wounded soldiers.

What no one realized at the time was how damaging too much radiation could be. Yes, they knew that radioactive materials needed to be handled with care because they could burn your skin, but they did not appreciate the fundamental damage it would do to the genetics of a person's cells.

Only in the 1920s did anyone start to draw a link between radiation and cancer, but it was too late for Marie. She already had cancer of the blood cells, leukaemia, and to this day her laboratory notebooks are locked away because they are too "hot" to handle.

Timeline

1867
Born in Warsaw, Poland, she was initially named Manya Sklodowska

1891
Leaves Poland to study physics at the Sorbonne in Paris

1893
Graduates at the top of her class

1894
Marries Pierre Curie

1896
Discovers that radioactivity is an atomic property of uranium

1897
Has her first child, Irène

1898
Discovers two radioactive elements and names them polonium and radium

1903
Awarded the Nobel Prize for Physics along with Pierre Curie and Henri Becquerel

1905
Gives birth to a second daughter, Eve

1906
Pierre Curie is killed when he steps in front of a horse-drawn carriage and his skull is crushed

1910
Marie publishes her work on radioactivity

1911
Marie is awarded the Nobel Prize for Chemistry

1914–1918
Helps to set up X-ray units during World War I

1918
Takes post as director of the Radium Institute in Paris

1934
Dies from leukaemia on July 4 near Sallanches, France

opposite *Marie Curie and her eldest daughter Irène Curie at work in their laboratory. These two French scientists are the only mother-daughter pair of Nobel Prize winners.*

With some scientists still unsure whether atoms existed, Ernest Rutherford stepped onto the scene and provided the final proof. At the same time, however, he showed that atoms were certainly not Democritus' idea of indivisible objects, nor Dalton's concept of immutable objects. Rutherford developed a new understanding that was so radical it required a new set of words. His view of the atom was that it consisted of a tiny, densely packed nucleus surrounded by orbiting electrons. The nucleus could emit particles and become the nucleus of a different element. With a few modifications, the concept has survived into the 21st century.

Ernest Rutherford

1871–1937

Acquaintances

— Joseph John (J.J.) Thompson (1856–1940)
— Frederick Soddy (1877–1956)
— Hans Geiger (1882–1945)
— Niels Bohr (1885–1962)
— James Chadwick (1891–1974)
— Robert Oppenheimer (1904–1967)

Not alchemy, but radioactivity

Ernest Rutherford's family emigrated from England to New Zealand before he was born. They ran a successful farm on the South Island near Nelson, and Rutherford grew up enjoying the open air and working alongside his 11 other siblings. Being a bright child, he won a scholarship to New Zealand's University of Canterbury. He then won another scholarship to study at Cambridge University, England, where he made the first successful wireless transmission over two miles, a feat that in itself would have placed him in the history books.

At Cambridge, Rutherford met with J. J. Thompson who encouraged him to start looking at the recently discovered X-ray—this would set the course of the rest of his life.

In the early years of his work, Rutherford found that all radioactive elements that he knew of emitted two kinds of radiation. One type carried a positive charge and the other a negative charge. He called them alpha and beta particles respectively. He also showed that the output of radiation from a sample of an element decreases exponentially with time. Any given sample of a known element takes a fixed time for the rate of emission of rays to decrease to one half of its initial value—it has a specific half-life.

In itself, this was fairly radical, but the establishment was just about to have a real shock. While working with Frederick Soddy in 1901 and 1902, Rutherford found that when a radioactive element emitted an alpha or beta particle it spontaneously turned into a different element. Many scientists poured scorn on the idea, comparing it to alchemy, but eventually the evidence was strong enough to provide convincing proof.

Firing shells through tissue paper

Moving to the University of Manchester, Rutherford teamed up with Hans Geiger, of Geiger counter fame. Together they set up a center to study radiation, and changed the face of physics. The story goes that Rutherford set a student to work on what he thought was a fairly unlikely set of experiments. He had been studying the way that alpha particles could scatter when they hit other objects and his initial results suggested that they were so fast and so powerful that nothing would stand in their way. The

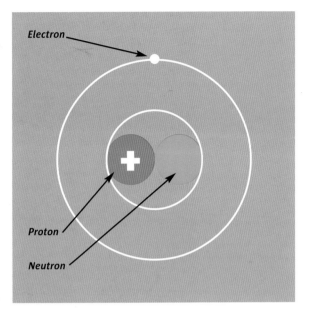

student set up a source of alpha particles in front of a sheet of gold foil and monitored the paths of the particles. There was every reason to assume that they would blast through any target, just as a 15-inch artillery shell would smash through tissue paper.

Then the remarkable happened. Some of the particles bounced straight back toward the source. Rutherford recounted that watching this was the most remarkable event that ever happened to him. It didn't take him long to conclude what must be occurring. An atom, he said, must consist of a minute, but dense nucleus, surrounded by a lot of space in which electrons orbit, much as planets orbit the Sun. The alpha particles that bounced back had taken a direct hit on the nucleus.

During World War I, Rutherford turned his attention to building systems for detecting submarines, but once hostilities were over he returned to his laboratory. This time he bombarded nitrogen gas with alpha particles and obtained oxygen and protons. He had deliberately split a non-radioactive atom and produced the first artificially induced nuclear reaction.

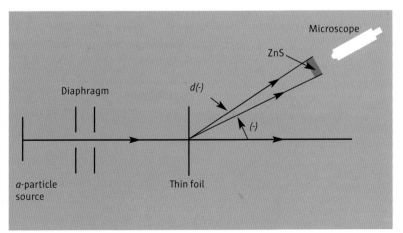

above *Hydrogen is the simplest atom with one proton and one neutron, plus one orbiting electron.*

left *When they fired alpha particles at a thin foil, Rutherford and his students were amazed to see that some were deflected—atoms must therefore have nuclei.*

There are two ways that you can influence science. You can be a great scientist, or you can be a facilitator of great science. Niels Bohr was both. In his work he brought an understanding of the atom that revealed how electrons could exist only in a few distinct orbits—rather than Rutherford's model, which left the electron free to orbit the nucleus at will. Bohr's quantization of the orbits explained why atoms emitted or absorbed specific frequencies of electromagnetic radiation. He turned Copenhagen into the world centre for theoretical physics research in the 1920s.

Niels Bohr

1885–1962

Acquaintances
—Ernest Rutherford (1871–1937)
—Albert Einstein (1879–1955)
—Max Born (1882–1970)
—Erwin Schrödinger (1887–1961)
—Linus Wolfgang Pauli (1900–1958)
—Werner Heisenberg (1901–1976)
—Paul Dirac (1902–1984)

Quantum orbits for electrons

Niels Bohr was born into a family of distinguished academics. His father was the professor of physiology at Copenhagen and his younger brother was a gifted mathematician. The distinction continued through the generations and Niels and his son Aage both won Nobel Prizes for Physics in 1922 and 1975 respectively.

Having completed his doctorate in Copenhagen, Bohr went to Cambridge to work with J.J. Thompson. The two didn't get on too well, as Thompson seemed not to have taken to Bohr's ideas, so Bohr quickly moved on to join Ernest Rutherford in Manchester. While there Bohr took another look at Rutherford's model of the atom, which envisaged negatively charged electrons orbiting a central positively charged nucleus.

The problem with the model was that there was nothing to keep the orbits of the electrons stable. That was unless you added some restraining features. Bohr suggested that the electrons would have to exist in one of a number of specific orbits, each orbit being defined by specific levels of energy. The electrons could move to higher level orbits if energy was added, or fall to lower ones if they gave out energy. Electrons could not, however, exist in between these definite steps.

This "quantized" theory of the electrons' orbits had two benefits. It explained why atoms always emitted or absorbed specific frequencies of electromagnetic radiation, and it provided an understanding of why atoms were stable.

In June 1912, Bohr wrote to his brother, "Perhaps I have found out a little about the structure of atoms. Don't talk about it to anyone, for otherwise I couldn't write to you about it so soon ... You understand that I may yet be wrong; for it hasn't been worked out fully yet (but I don't think it is wrong)." It wasn't. Working from his model, in 1913 Bohr calculated what the emission and absorption spectra of hydrogen atoms should be. Gratifyingly it fitted with experimental data.

Copenhagen conversations

In 1913 Bohr took a bold step and wrote to the Danish Department of Educational Affairs calling on them to create a position of professor in theoretical physics and to appoint him to the post. In 1921, with much debate and one world war over, the Danish Academy of Sciences opened the Institute of Theoretical Physics in Copenhagen with Bohr as its director.

The institute turned Copenhagen into one of the key intellectual centers of the world. During the 1920s and 1930s numerous physicists, chemists, and biologists flocked to speak with Bohr and join in the feast of academic debate. The guest list reads like a roll of honor, with some of the most famous being Erwin Schrödinger, Linus Wolfgang Pauli, Werner Heisenberg, Paul Dirac, and Max Born. Bohr showed a talent for spotting people's gifts and finding and supporting young researchers.

Being passionate about his subject, Bohr could become totally absorbed. Once, he and Einstein became so engrossed in conversation while taking a short streetcar ride to the Institute that they missed their stop, and apparently continued for some time before they spotted the error. Getting out, they caught a car going back again. Once more they traveled too far. They repeated the process a few more times before eventually disembarking at the Institute, their intended destination. Einstein once said of Bohr, "He utters his opinions like one who perpetually gropes and never like one who believes to possess the definite truth." From Einstein that was high praise.

Bohr was also determined to pursue an issue until it was fully worked out, and on one occasion refused to delay a crucial discussion with Schrödinger simply because his fellow scientist was sick in bed with a fever. Instead, he sat on the edge of Schrödinger's bed pummeling him with questions while Margrethe Bohr, his wife, nursed the ailing Austrian and served tea.

Disturbed by history

Word War II led to the break-up of the club. As his mother had been a Jew, Bohr was forced to flee from Denmark in 1943 by taking a fishing boat to Sweden. From there he flew to England in the bomb bay of a Mosquito aircraft, and then went on to Los Alamos to work on the atomic bomb project.

Afraid of losing his Nobel Prize medal, or having it forcibly taken from him, he had dissolved it in acid before leaving Copenhagen and left the bottle behind. Not realizing its value, no one touched it, and on his return to Copenhagen after the war he recovered the metal and had it recast.

Niels Bohr died of a heart attack in 1962.

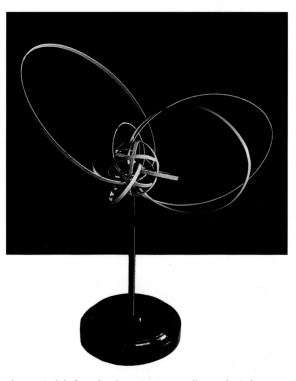

above *Model of an aluminum atom according to the Bohr theory of the structure of atoms.*

Many people live full lives. Some live truly active lives. A few people seem to live many lives. Linus Pauling was one such person, and it is fitting that this was recognized in his two Nobel Prizes. The first was awarded for his pioneering and imaginative work in chemistry. The second, for his campaigning in the name of peace. His life was full of contrasts. He was decorated by the U.S. president as a war hero, but was also persecuted by the U.S. government for being anti-American. As one of the most influential and controversial figures of the 20th century, he is the only person so far to receive two unshared Nobel Prizes.

Linus Pauling

1901–1994

Acquaintances
- Arnold Sommerfeld (1868–1951)
- Richard C. Tolman (1881–1948)
- Niels Bohr (1885–1962)
- Erwin Schrödinger (1887–1961)
- Werner Heisenberg (1901–1976)
- Robert Oppenheimer (1904–1967)
- Max Delbrück (1906–1981)
- Francis Crick (1916–)
- James Watson (1928–)

Science lying down

Linus Pauling used to think about scientific dilemmas just before turning off the light for the night. He claimed that he had trained his subconscious to work while he rested, and found that if he slept on an issue the solution could at times "jump up" and hit him in the form of an instinctive feeling.

The process seems to have been highly successful. In 1931, Pauling wrote his most important scientific paper, entitled "The Nature of the Chemical Bond." In this he suggested that in order to create stronger bonds, atoms change the shapes of their waves into petal-shapes, a concept that became known as the "hybridization of orbitals." This allowed him to develop six key rules that enable scientists to explain and predict chemical structure. Three of them are mathematical rules relating to the way electrons behave within bonds, and three relate to the orientation of the orbitals in which the electrons move and the relative positioning of the atomic nuclei.

One aspect of the revolution he brought to chemistry was to lift structures off the flat two-dimensional page and insist on talking about them in terms of their three-dimensional space. Like most of his work, it was controversial at first, but was rapidly accepted and incorporated into mainstream thought.

Pauling had been brought up in the era of a quantum understanding of science, and was convinced that it was the best way of explaining the behavior of atoms. One problem that needed resolving was the distance between particular atoms when they joined together. For example, carbon was known to form four bonds, while oxygen can form two. It seems obvious that in a molecule of carbon dioxide, which is made of one carbon and two oxygen atoms, two of carbon's bonds will be devoted to each oxygen. From the well-established calculations this told scientists that the distance between the carbon and the oxygen atoms should be 1.22 x 10-10m. In fact, analysis of carbon dioxide showed that it was 1.16 Angstroms. The bond was stronger and hence shorter than it should have been.

The explanation, Pauling said, was that the bonds within carbon dioxide are constantly resonating between two alternatives. In one position, carbon makes three bonds with one of the oxygen molecules and has only one bond with the other, and then the situation is reversed. It might seem like a strange arrangement, but it is perfectly within the rules of quantum understanding.

Molecular biology

With the world at war, Pauling began to shift his attention to chemical biology, and in 1934 he applied to the Rockefeller Foundation for a three-year grant to study hemoglobin, the protein that transports oxygen in blood. Through the late 1930s and 1940s he worked on developing plasma substitutes that could be used to help injured soldiers, as blood loss from wounds was a major cause of death on the battlefield. His wartime efforts were rewarded when in 1948 President Harry Truman gave him the Presidential Medal of Merit.

above *A mushroom cloud of water and radioactive material. Campaigning for a non-nuclear future got Pauling into trouble.*

In 1940, when working with Max Delbrück, he proposed that the body's immune system has antibodies that lock onto invading particles, such as bacteria and viruses, by forming a chemical bond. The bond was possible, because the antibodies had specially shaped chemicals on the surface that fitted somewhat jigsaw-like into the surface of the foreign material.

In seeing that the shape of biological chemicals was important, Pauling became intrigued by sickle cell anemia. This is a disease in which blood fails to transport enough oxygen and the red cells are badly deformed. With colleague Harvey Itano he showed that the disease was caused by a genetic abnormality, and he referred to it as a molecular disease. In doing so he triggered the train of thought that has led to an explosion in genetic research, some of which is aimed at curing genetic disease.

No more war

Pauling's understanding of science and biochemistry heightened his concern about nuclear warfare. Inspired and goaded by his wife Ava Helen, he researched the issue thoroughly, looking into the impact of nuclear fallout on human health. His findings were shocking, indicating that everyone in the world was likely to have a shortened life span as a result of the nuclear bomb tests. He rejected the concept that radiation around the world was acceptable as long as it was kept below certain levels.

As part of his campaign he collected a petition containing the names of 11,000 scientists from 49 countries who all called for a ban. This was presented to the United Nations, and partly led to Pauling been summoned before a senate sub-committee on suspicion of being a communist and threatened with imprisonment when he refused to give the names of sources of some information. One consequence was that the U.S. State Department refused him a passport, an act that became an embarrassment when in 1954 he was awarded the Nobel Prize for Chemistry. His passport arrived two weeks before the ceremony.

In 1958 he published his views in a book, *No More War*, and his efforts were instrumental in the signing of the partial nuclear test ban treaty in 1963. On the same day the Nobel Prize committee awarded him the 1962 Peace Prize. He died on his coastal ranch in California on August 19, 1994.

Almost all of the molecules that make up living things are large and contain carbon. These so-called organic compounds have complicated structures, but if you want to make sense of the properties and reactions of one of the molecules you need to know its structure. This is even more important should you ever want to build synthetic versions of the compound by joining together simpler compounds. X-rays are unique in that their wavelength is about the length of bonds inside molecules, a feature that allowed Dorothy Hodgkin (née Crowfoot) to employ this part of the electromagnetic spectrum to determine the structure of many biologically important compounds.

Dorothy Crowfoot Hodgkin

1910–1994

Acquaintances
— John Desmond Bernal (1901–1971)

X-ray diffraction

In the latter part of the 19th century, chemists started calculating the shapes of some carbon-containing compounds. They built models by making assumptions about what bonds would form. This worked quite well with smaller molecules, but the larger ones were too complex.

In 1914 Max von Laue was awarded a Nobel Prize for his discovery of "the diffraction of X-rays by crystals." This opened a new possibility, and a year later father and son William H. Bragg and William L. Bragg shared the 1915 Nobel Prize for using X-ray diffraction to determine how the atoms of a compound are positioned relative to each other.

The scene was set for people to start making sense of the three-dimensional structure of compounds. The problem was that some of the most useful and therefore most valuable compounds turned out to be highly complex, with each molecule containing many hundreds of atoms, each held in a precise location.

By the time Hodgkin finished her first degree she was uncertain whether to pursue studies in antiquities or in crystals. Both subjects fascinated her and at first she hoped to solve the dilemma by using X-ray crystallography to help analyze ancient samples. Her tutor, F.M. Brewer, persuaded her to concentrate on crystals, and launched her career.

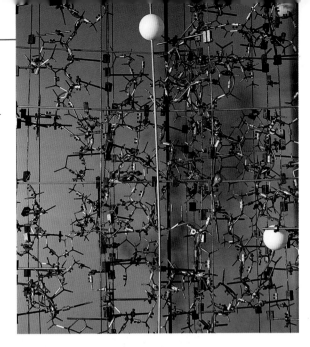

With Brewer's advice she moved to Cambridge to work with John Desmond Bernal. Bernal was developing techniques of X-ray crystallography. The idea is that when X-rays hit a crystallized molecule, the electrons surrounding each atom cause the beam to bend. Because there are many atoms the result is that when the X-rays exit the crystal and fall onto a photographic plate, they produce a series of light and dark patches. Measuring the intensity and relative position of each patch indicates the relative positions of atoms within the crystal.

In a world that had not yet invented computers, the mathematics required to solve complex compounds was formidable, but together Hodgkin and Bernal produced the first diffraction patterns for proteins.

Determining structure

The first real success came from a crystallized form of pepsin, using crystals which had been produced by John Philpot in Uppsala, Sweden. One of their early discoveries was that if you wanted to see the structure of a protein, you had to keep the crystal in the liquid that it would normally exist in. Drying it or placing it in another solution radically altered the protein's shape, making any observations worthless.

At the end of her Ph.D., Hodgkin returned to Oxford, and among other things started work on insulin. This was a highly complex task and would take her 34 years to complete. The reason for part of her success, apart from brute perseverance, was that she developed ways of building crystals in which the molecules had specific atoms replaced by zinc. These crystals revealed clues about the structure of the naturally occurring protein.

During this research penicillin arrived on the scene. Clearly this was an important compound. It was, however, difficult to produce naturally, and before it could be manufactured someone was going to have to work out its structure. Hodgkin found that along with its intriguing function in killing bacteria, it had an unusual ring feature, now known as the Beta-lactam structure.

In 1955 she took the first X-ray diffraction photos of cyanocobalamin crystals, more commonly called vitamin B12. This molecule proved to be four times the size of penicillin and again contained a core ring structure. This porphryin ring contained cobalt in the center.

By this stage Hodgkin was pioneering the use of computers to assist in the calculations, and collaboration with Kenneth Trueblood at the University of California in Los Angeles gave her access to state-of-the-art processors.

Hodgkin's work was considered to be so important that she became the first woman since Florence Nightingale to be called by the Queen into the Order of Merit.

top right *Hodgkin's model of pig insulin, c. 1967. Insulin is a hormone that controls the body's sugar metabolism. This model shows the complicated structure of insulin. Understanding this structure has allowed pharmaceutical companies to produce human-type insulin.*

Timeline

1910
Born on May 12 in Cairo where her father, John Crowfoot, was working in the Egyptian Education Service

1928–1932
Studies for her first degree at Somerville College, Oxford

1937
Obtains her Ph.D. at Cambridge University

1932–1936
Works with John Desmond Bernal, who reinforced her interest in structural biochemistry

1934
Bernal and Crowfoot report the diffraction pattern of pepsin

1936
Crowfoot becomes a tutor at Somerville College, Oxford

1937
Marries Thomas Hodgkin, and the couple subsequently have two sons and a daughter

1946
She takes an appointment as a university lecturer and demonstrator

1956
Produces a three-dimensional analysis of vitamin B12

1960
Becomes the Wolfson Research Professor of the Royal Society

1964
Receives the Nobel Prize for Chemistry, principally for her determination of the structure of vitamin B12

1965
Is given the Order of Merit, the U.K.'s highest royal order

1969
Reports the structure of insulin

1971
Becomes a member of the United States National Academy of Sciences

1994
Dies on July 29

Quantum theory of electromagnetic fields was a puzzle for the scientists when Richard Feynman was in college. One of the problems was that for the theory to work electrons needed to have an infinite energy of interaction and an infinite number of degrees of freedom. Feynman solved it by producing his version of quantum electrodynamics, a new generation of quantum theory. His inquiring mind and ruthless independence led him to take up a hobby of safe-cracking while working at the top secret Los Alamos laboratories, and made him an outspoken member of the commission called to investigate the loss of the space shuttle Challenger.

Richard Feynman

1918–1988

Acquaintances
—Sin-Itiro Tomonaga (1906–1979)
—Hans Bethe (1906–)
—Julian Schwinger (1918–1994)

above *The launch of the ill-fated space shuttle Challenger mission 51- L. Seventy three seconds after lift-off the spacecraft exploded killing all seven crew members. It is thought that O-rings in a seal on the right hand solid rocket booster failed and the subsequent leak of fuel resulted in the explosion.*

Infinite problems at the limits

Quantum electrodynamics was the successor to quantum mechanics. But, while being an improvement, it still had problems. The concept appeared to be fundamentally correct, but when physicists tried to calculate answers, they ran into complicated equations that were hard to solve. It was possible to get good approximate answers, but when they tried to push the equations hard and get detailed results, problems set in. Infinite quantities started to crop up, and this couldn't be real.

The dilemma got larger when Willis Eugene Lamb performed some experiments on hydrogen atoms that showed the equations had serious limitations. The issue had to be solved.

At this point, Feynman was working with Hans Bethe, the director of theoretical physics for the atomic bomb project at Los Alamos. In great excitement one day, Bethe showed Feynman a way of getting around the problem. Feynman was interested, but not impressed. He thought that Bethe was pointing the way forward, but had done so by creating even more complexity.

A month or so later, Feynman returned with his lean and clean way of calculating quantum electrodynamics. It may not have changed the world, but it set quantum theory on its soundest ever footing.

Not so safe

When called to work on the Manhattan Project at Los Alamos he developed a new pastime. Much to the authorities' anxiety he set about learning how to open locked safes. Sometimes by listening to the tiny movements of the mechanism, at others by taking guesses about the physical constants that the maker and user of the safe had chosen to use. It's a capability that he reportedly enjoyed passing on to future students, many of whom have left Caltech with this additional skill.

Challenger disaster report

On January 28, 1986, the space shuttle Challenger blew up in full view of the world's media, just after take-off. It was a U.S. national disaster and a public relations nightmare for NASA. The President of the United States stepped in and ordered an independent inquiry into the accident and NASA invited Feynman to join the commission. They probably regretted the decision when he showed the extent of his determination to remain an independent observer and, to that end, published his own appendix to the report.

The official report pointed to rubber sealing rings that were vital parts of the rocket motor's safety. Feynman publicly demonstrated how rubber, when cold, becomes brittle. The report also leveled criticism at NASA's system of management. Feynman added to this by poking fun at the management's strange use of statistics that attempted to show that the shuttle was safe. He gave two reasons why he thought NASA took this approach. "One reason for this may be an attempt to assure the government of NASA perfection and success in order to ensure the supply of funds. The other may be that they sincerely believed it to be true, demonstrating an almost incredible lack of communication between themselves and their working engineers ... For a successful technology, reality must take precedence over public relations, for nature cannot be fooled."

Timeline

1918
Born May 11 in Queens, New York
1936
Enters Massachusetts Institute of Technology to study physics
1939
Moves to Princeton for his Ph.D.
1940
Marries Arline Greenbaum
1942
Moves to Los Alamos, N.M.
1945
Arline dies of tuberculosis
1950
Moves to the California Institute of Technology
1952
Marries Mary Louise Bell who he divorces in 1956
1960
Marries Gweneth Howarth. They have a son, Carl, and adopt a daughter, Michelle
1961–1963
Gives a set of lectures that are recorded and published in a series of three books: *The Feynman Lectures on Physics*
1965
Shares the Nobel Prize for Physics with Julian Schwinger and Sin-Itiro Tomonaga
1986
Joins Rogers commission to investigate shuttle disaster
1988
Dies on February 15 in Los Angeles, California

Charles Lyell

1797–1875

Acquaintances
- George Cuvier (1769–1832)
- Alexander von Humboldt (1769–1859)
- William Buckland (1784–1856)
- Roderick Impey Murchison (1792–1871)
- Charles Darwin (1809–1882)
- Alfred Wallace (1823–1913)
- James Hutton (1726–1797)

As the 19th century dawned, most people believed that a few major events had shaped the Earth, one of which was Noah's great biblical flood. In between these catastrophic incidents, the Earth had remained unchanged. Funding his work from his family fortune and from the publication of his books, Charles Lyell traveled the world studying rock formations. His observations led him to replace catastrophe theory with uniformitarianism, which proposed that the Earth changed gradually as constantly-present forces acted on it. He also attributed ages to rock strata, by looking at the fossils that they contained. This introduced a new way of studying the Earth and led to modern geology.

Catastrophe versus uniformitarianism

When Charles Lyell looked at the Earth he was dissatisfied with the explanations that people gave for the origins of its physical structures. The basic concept was that God had created the world and that there had been a few major incidents that had shaped it, but other than that it hadn't changed appreciably.

While studying at Oxford, Lyell was intrigued by lectures given by William Buckland, a clergyman who was studying the Earth. Buckland introduced him to the work of James Hutton, a scientist who proposed that the Earth changed gradually and more or less constantly. Hutton, however, had little evidence to support the concept. To get the evidence you would need not only an interest in the subject but also the ability to travel, and that would be costly. Lyell had both. He had inherited plenty of money and had an inquiring mind, a mind that was prepared to think "outside the box."

In an essay entitled The Progress of Geology, Lyell explains his reasoning. He begins by using an analogy to attack the concept that the Earth has existed for only a few thousand years. Say, he suggests, you visited Egypt with a belief that the banks of the Nile had not been populated before the 19th century — in other words, the present time — "and that their faith in this dogma was as difficult to shake as the opinion of our ancestors, that the Earth was never the abode of living beings until the creation of the present continents, and of the species now existing ..." How, he asks, would they explain the pyramids, obelisks, statues, ruined temples, and the mummies?

The options he suggests are that these are the actions of some super-human powers, or that objects like mummies had been "generated by some plastic virtue residing in the interior of the earth." The other option is to question the initial dogma, and allow the possibility that humans have lived by the Nile for many more years.

So too, said Lyell, looking at the Earth demands that we examine the possibility that it has existed for millions of years. This allows for the idea that it could have been shaped gradually, and removes the necessity to explain everything in terms of violent, short-lived acts. Instead, uniform processes could act on the Earth, slowly creating change and, what's more, if you looked for them you could see the process occurring all around the world every day. While in the United States, he estimated how quickly the Niagara Falls were eroding their riverbed and moving back upstream. He also calculated the average rate at which mud and other alluvial matter collected in the Mississippi delta, and studied the way that vegetative matter built up in the "Great Dismal Swamp" of Virginia. Later, he used this information to speculate about how coal beds form.

Hidden secrets

The first and greatest difficulty, said Lyell, is that we fail to take into account the fact that we can only observe about one quarter of the surface of the planet. The rest, he pointed out, lies beneath the seas and oceans. To that he adds the problem of not understanding where lava comes from, or what effect it has as it forces its way up through rock strata on the way to creating a volcano. He was also concerned by the lack of knowledge about underground rivers and "reservoirs of liquid matter" that he presumed existed beneath the Earth's surface.

Eon	Era	Period	Epoch	Abs. Age (ma)	Age of	Events
	Cenozoic	Quaternary	Holocene	0-2	Mammals	Humans
			Pleistocene			Scabland Floods
		Neogene	Pliocene	2-5		
			Miocene	5-24		Columbia Basalts
		Tertiary	Oligocene	24-37		
		Paleogene	Eocene	37-58		
			Paleocene	58-66		Extinction of Dinosaurs
Proterozoic	Mesozoic	Cretaceous		66-144	Reptiles	Flowering Plants
		Jurassic		144-208		1st Birds/Mammals
		Triassic		208-245		1st Dinosaurs
	Paleozoic	Permian		245-286	Amphibians	Extinction of Trilobites
		Carboniferous – Pennsylvanian		286-320		1st Reptiles
		Carboniferous – Mississippian		320-360		Large Primitive Trees
		Devonian		360-408	Fishes	1st Amphibians
		Silurian		408-438		1st Land Plants
		Ordovician		438-505	Invertebrates	1st Fish
		Cambrian		505-570		1st Shells, Trilobites
Proterozoic		Precambrian		570-2,500		1st Multicelled organisms
Archean				2,500-3,800		1st One-celled organisms
Hadean				3,800-4,600		Origin of the Earth

above *Lyell started a chain of thought that has now generated a complex understanding of the Earth's history, allowing it to be divided into discrete eons, eras, periods, and epochs.*

Four periods

The subtitle of his major book, *The Principles of Geology*, indicates the thrust of his work: *An Attempt to Explain the Former Changes of the Earth's Surface by Reference to Causes Now in Operation.* In 1828 he traveled with pioneering geologist Roderick Murchison through the south of France looking at the volcanic district of Auvergne and the rock formations of Aix-en-Provence. He then continued alone to Italy where he devised a system for dating rocks, based on the fossilized marine shells he found encased in each layer.

If the different ages of the rocks were going to be adopted for general use they would need titles, and in consultation with Cambridge mineralogist William Whewell he came up with four names that are still in common use — Eocene referred to the youngest rocks, with Miocene, Pliocene and Pleistocene, getting progressively older. He also introduced the idea of metamorphic rocks, rocks that can change their form when exposed to high temperature and pressure.

His book was so successful that he produced seven editions of *The Principles of Geology*, updating each new edition with his latest thoughts and observations. Like developing layers of sediment, the sequence of the editions therefore gives an intriguing chronological insight into the development of his thoughts.

Acquainted with Darwin

There is plenty of evidence that Darwin was influenced by Lyell, but there is less that the opposite occurred. Indeed, Lyell was deeply troubled by Darwin's concept of natural selection, as he felt that this conflicted with his own religious beliefs. It appears that he had reluctantly accepted at least some of Darwin's ideas by the time he died. Darwin was still highly complimentary: "The greatest merit of the *Principles* was that it altered the whole tone of one's mind, and therefore that, when seeing a thing never seen by Lyell, one yet saw it through his eyes."

One aspect of his work that Lyell did get wrong was his concept that geological, and therefore biological, history was cyclical. He thought that there were slow but repeating patterns in climates, that would be accompanied by species best suited to each set of conditions. This led him to conclude that "huge Iguanodon might reappear in the woods, and Ichthyosaurs in the sea, while Pterodactyls might flit again through umbrageous groves of tree ferns."

While Lyell had destroyed one major dogma, his adherence to other ideas prevented geology moving forward as fast as it might.

Timeline

1797
Born at Kinnordy House, Forfarshire, the center of his grandfather's estate in Scotland on November 14. He was the eldest of 10 children

1816
Enters Exeter College, Oxford, where he is captivated by the lectures of geologist and clergyman William Buckland

1819
After getting his B.A. degree Lyell entered Lincoln's Inn, London, to study law. In the same year he is elected to the Linnean and Geographical Societies, and presents his first paper, entitled "On a recent formation of freshwater limestone in Forfarshire," to the Geographical Society

1823
Tours France with George Cuvier and Alexander von Humboldt

1824
Tours Scotland with Buckland

1825
After a delay caused by poor eyesight, Lyell is called to the Bar and serves on the "Western Circuit" for two years

1826
Elected a fellow of the Royal Society

1827
Formally abandons the legal profession to devote his efforts to geology

1830
Publishes first volume of *The Principles of Geology*

1831
Marries Mary Horner

1832
Publishes second volume of *The Principles of Geology*

1834
Tours Denmark and Sweden

1838
Publishes *Elements of Geology*

1841
Travels in the United States, Canada, and Nova Scotia

1845
Makes a second trip to North America

1848
Receives a knighthood from Queen Victoria at Balmoral

1858
Visits Sicily

1863
Publishes *The Antiquity of Man*

1864
Becomes a baron

1875
Dies in London on February 22, and is buried in Westminster Abbey

Introducing a new idea is never easy, particularly if your theories affect a different academic discipline from the one you are trained in. This was a key problem for Alfred Wegener. After studying for a Ph.D. in astronomy he started to concentrate on monitoring weather patterns in extreme climates like Greenland. In his spare time, however, he was fascinated by the possibility that America and Africa had at one time been joined. He proposed the concept that the two great continents had drifted apart, and in so doing he led to the understanding that the Earth's surface is made of a number of separate plates.

Alfred Wegener

1880-1930

Acquaintances
— Wladimir Köppen (1846–1940)
— Johann Peter Koch (1870–1920)
— Else Wegener (1892–1992)

Drifting continents

Just as people were getting used to the idea that the world was in a state of slow, but continual, change, along came Alfred Wegener with a theory that suggested that the forces at work could be greater than anyone had dreamed of. He had been wounded fighting in World War I and, while he lay in bed recovering, he started to reflect on some reading he had done in the library at the University of Marburg three years earlier.

While carrying out research into weather conditions in 1911, he had come across a scientific paper that listed fossils of identical plants and animals found on opposite sides of the Atlantic. This had intrigued him and he decided that once he was sufficiently fit he would start to look for other examples of situations where similar fossils were separated by seas or oceans.

The existing theory was that the two continents must at some point have been linked by a land-bridge, which had over the years been eroded or sunk and consequently replaced by the ocean. But Wegener did not find the theory of sinking land-bridges convincing. Instead, he looked hard at the shapes of the African and South American coastlines on either side of the Atlantic. Remarkably, he found that if you were to cut the continents out from a map, they would effectively fit together.

He wasn't the first person to spot this. The Flemish map-maker Abraham Ortel Ortelius had in 1596 suggested that the continents had been ripped apart by floods and earthquakes. In 1620, the English philosopher and statesman Francis Bacon also mentioned that the continents could fit together as if they were cut from the same mold, and in 1858 the French scientist Antonio Snider-Pelligrini showed in his book *Creation and its Mysteries Revealed*, how the continents fitted together before Noah's flood.

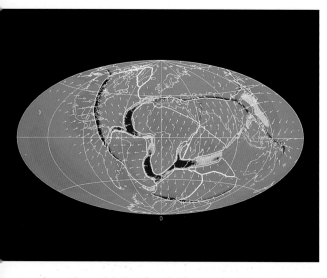

above *The models that describe the ways that the earth has constantly changed show that the landmasses were originally grouped into one large continent. Wegener called it Pangenia.*

Wegener had spoken to people about the idea in 1912, but was not taken seriously. Most geologists had dismissed him as a mere meteorologist and outsider who was meddling in their field. Now, during his prolonged convalescence, he had time to read and gather evidence.

He soon found that the Appalachian Mountains of eastern North America shared remarkable similarities with the Scottish Highlands, and that the distinctive form of strata in the Karroo system of South Africa were identical to those of Santa Catarina in Brazil. In addition, fossils were turning up in the wrong places. Fossilized ferns and cycads were found on the Arctic island of Spitzbergen, even though the plants grow only in tropical climates. Clearly, thought Wegener, the rock masses have migrated.

By 1915, Wegener had collected enough evidence to publish a book: *The Origins of Continents and Oceans*. In this he proposed that some 300 million years ago there was one continent. He called it Pangaea, from the Greek for "all the earth." Since then it had fragmented, and the pieces were drifting apart.

Keeping it in the family

Wegener died while returning from a research mission to get supplies to a base on Greenland. His wife, Else, was not about to let his memory die. She had always taken a vigorous part in his scientific work, and she recorded the events of that mission in her book, *Alfred Wegener's Last Journey to Greenland*. In addition, Wegener's brother took over his professorship in Graz.

Else Wegener also wrote a biography of her father, Wladimir Köppen: *A Life of Scholarship for Meteorology*. This told the story of his contribution to an understanding of how the Earth can be divided into different climatic regions. He was not the first person to have tried—Aristotle had tackled the subject and developed the idea of zones, each of which was defined simply by its distance from the Equator. It was a good concept, but there was not enough data to get it right.

Aristotle believed that the region on either side of the Equator would be too hot for living organisms and named it the Torrid zone. The Temperate zone was occupied by regions like Europe and North Africa, and was, he thought, the place that contained all life. Further from the Equator was the Frigid zone, a place too cold to support life.

Some 2,000 years later, Wegener's father-in-law, Köppen, a German meteorologist, climatologist, and amateur botanist came up with an alternative system. This has stood the test of time and forms the basis for all current climate maps.

In 1928, along with student Rudolph Geiger, he introduced a new map. With more recent modifications, this classifies the world into one of six climates: Tropical humid, Dry, Mild mid-latitude, Severe mid-latitude, Polar and Highland. The aim is to provide a guide to the general climate of the regions, and the map will need constant modification if predicted climate change takes place.

Lacking a motivating force

If he thought that presenting a mass of evidence would quiet his critics, Wegener was wrong. Rollin Chamberlin of the University of Chicago reflected the general opinion, when he said, "Wegener's hypothesis in general is of the footloose type, in that it takes considerable liberty with our globe, and is less bound by restrictions or tied down by awkward, ugly facts than most of its rival theories."

Part of Wegener's problem in facing his critics was that he could produce no mechanism that looked feasible. After all, shifting continents demanded huge, and arguably inconceivable, quantities of energy. His best guess was that the continents had ploughed through the Earth's crust like an icebreaker moving through the ice sheets. He speculated that this movement was generated by centrifugal forces of the spinning Earth, assisted by tides.

Quick calculations, however, showed that the Earth's rotation was too slow to have any appreciable effect, and if it did spin around fast enough it would be ripped apart. In addition, there was every reason to assume that the shape of the continents would be radically altered by the process, and would no longer look as if they had once fitted together.

From floats to plates

Right up to and beyond Wegener's death, most scientists were unconvinced. But then, in the 1950s, new technologies allowed people to start studying ocean floors. The discovery of a ridge running down the middle of the Atlantic Ocean produced the missing segment in Wegener's theoretical jigsaw. The ridge is caused by molten rock forcing up from beneath the Earth's crust. This creates a new ocean floor and pushes the continents apart.

The model then changed from Wegener's idea of continents that float and drift around the surface of the Earth, which he called "continental drift," to an understanding that the surface of the Earth is made of plates. In some places, these plates are being forced apart, in others they are crashing into each other and buckling to form mountains, and in still other places, one plate is being forced beneath another. This is the currently accepted concept of plate tectonics.

Timeline

1880
Born on November 1

1904
Receives his Ph.D. in astronomy from the University of Berlin

1906
Joins an expedition to Greenland to study polar air circulation and accepts a post as tutor at the University of Marburg, Germany

1911
Starts to look for evidence that the Americas had at one time been physically joined to Africa and had been pushed apart

1912–1913
Visits Greenland again, crossing the ice cap of Greenland from east to west with the Danish explorer Johann Peter Koch between Dove Bay and Upernavik. This is the first wintering of Europeans in Greenland

1912
Marries Else Köppen, the daughter of Germany's leading meteorologist

1914
First daughter, Hilde, is born

1914
Drafted into the German army, but is wounded and serves the rest of the war in the army weather-forecasting service

1915
Publishes the first edition of *The Origins of Continents and Oceans*

1918
Second daughter, Käthe, is born

1919
Becomes head of the department for theoretical meteorology at the German Hydrographic Office, Hamburg, where he takes over from his father-in-law

1920
Third daughter, Charlotte, is born

1924
Accepts a carefully created post of professor of meteorology and geophysics at the University of Graz, Austria, and accepts Austrian citizenship

1930
Makes fateful expedition to Greenland and dies while trying to reach the camp on the west coast of Greenland. The exact date of death is not known but was a day or so after his fiftieth birthday

1931
Exploration stays in the family as Alfred Wegener's brother, Kurt Wegener (1878–1964), continues leading the expedition, and takes over his professorship in Graz

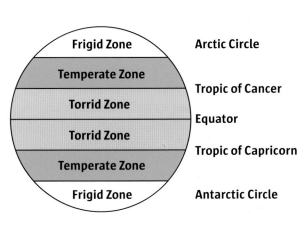

Frigid Zone — Arctic Circle
Temperate Zone
Torrid Zone — Tropic of Cancer
Equator
Torrid Zone
Temperate Zone — Tropic of Capricorn
Frigid Zone — Antarctic Circle

left *The ancient Greek scholar Aristotle divided the Earth into three types of climatic zone. He believed that life could only survive in the temperate zone—the torrid zone would be too hot and the frigid zone too cold.*

Ancient Greek thinkers often held the view that spirits or gods inhabited every physical part of the world. To that extent the world was alive. More recently, science has divided the world into animate and inanimate categories, and philosophers have developed theories that see individual organisms developing as they aggressively compete for survival. Into this, James Lovelock introduced his Gaia hypothesis, suggesting that it is more appropriate to see the entire world as a single living organism. According to Lovelock, life has evolved so that it maintains the Earth's stability and health. Like all good theories, it has its opponents.

1919–

Acquaintances
—Lynn Margulis (1938–)
—William Golding (1911–1993)

James Lovelock

When looking for life

While working as a chemist at the Jet Propulsion Laboratory in Pasadena, California, in the 1950s, James Lovelock invented a detector that was capable of tracing extremely small amounts of chemicals in gases. He used this electron capture detector to look for the presence of pesticides around the world, and came to the shocking conclusion that these chemicals had spread to every region. DDT, for example, appeared in penguins in Antarctica and mothers' milk in America. The data led to a world-wide effort to ban DDT, and fuelled a growing environmental campaign.

Impressed with his inventive skills, NASA asked Lovelock to join their Mars Viking project. The authorities were unaware that this would trigger a train of thought that would lead to us looking differently at our own planet.

If he was standing on Mars, Lovelock asked himself, how could he tell that there was life on Earth? His answer was that without life, the atmosphere of a planet would fall into a chemically-dead state of equilibrium. Planets inhabited by living organisms would, however, have living dynamic atmospheres.

He pointed out that the Martian atmosphere consisted of 95.3% carbon dioxide, 2.7% nitrogen and 1.6% argon, and only traces of oxygen and water. This had all the signs of a dead atmosphere. The obvious conclusion was that there was no life on the planet. In contrast the Earth's atmosphere contains 77% nitrogen, 21% oxygen, and only traces of carbon dioxide, methane, and argon. This mixture is far from a simple chemical equilibrium, and is a sign that some active process is causing it to exist.

Working with American microbiologist Lynn Margulis, Lovelock concluded that microbes, plants, and animals actively use the energy from sunlight to create an atmosphere that is suitable to sustain life. On this basis it is inappropriate to think of the Earth as an inert vehicle that transports living plants and animals through space. Instead, the entire planet should be seen as an integrated living bio-system.

Seeking a name

To an extent this was not a new way of looking at planet Earth, because it had echoes of some of the patterns of thought used in Ancient Greece. While on a walk one day with his neighbor, the novelist William Golding, best known for writing *Lord of the Flies*, Lovelock asked his advice about a name for his theory. Golding recommended Gaia, after the Greek goddess who drew the living world from chaos.

According to Lovelock, the name has the advantage of making the concept more acceptable to people who haven't studied science. Its disadvantage though, is that at times people have thought he intended to indicate that Gaia was a form of guiding intelligence — a concept that he denies.

Gaia at work

Lovelock sees the Earth as more than a ball of wet rock hurtling through space, saying there is evidence that life not only adapts to its environment, but also changes the environment to suit its own purposes.

Gaia theory also maintains that natural selection favors organisms that leave their environment in a better condition than they found it. As a consequence, living systems evolve so that they work together. In *Gaia: a New Look at Life on Earth*, he says that "the entire range of living matter on Earth from whales to viruses and from oaks to algae could be regarded as constituting a single living entity capable of maintaining the Earth's atmosphere to suit its overall needs." Gaia describes the control process by which living organisms create an optimal physical and chemical environment for their own existence.

The theory has many opponents because it runs counter to the thinking of Darwinian evolution, where organisms adapt to environments that they can't influence and compete for their independent survival. Lovelock is always quick to point out that the history of science shows how it takes time for new ideas to accumulate sufficient data before they gain widespread acceptance.

The theory is under intense scrutiny, but if nothing else it has already spawned a large interest in humanity's interaction with its environment.

Timeline

1919
Born in Letchworth Garden City, England, on July 26
1941
Graduates from Manchester University with a degree in chemistry and takes a post at the National Institute of Medical Research in London
1946–1951
Works at the Common Cold Research Unit at Harvard Hospital, Salisbury
1948
Receives a Ph.D. from the London School of Hygiene and Tropical Medicine
1957
Plays a critical role in inventing the electron capture detector, one of the most sensitive detectors of chemicals in existence
1959
Is awarded a D.Sc. in biophysics from the University of London
1961
Becomes the professor of chemistry at Baylor University College of Medicine in Houston, Texas
1964
During this year he resigns his position and works for the rest of his career as an independent scientist
1974
Elected a fellow of the Royal Society
1979
Publishes *Gaia: a New Look at Life on Earth*
1986–1990
President of the Marine Biological Association, Plymouth
1988
Publishes *The Ages of Gaia*
1991
Publishes *Gaia: the Practical Science of Planetary Medicine*

Edwin Hubble

1889–1953

Acquaintances
—Albert Einstein (1879–1955)
—Harlow Shapely (1885–1972)
—Charlie Chaplin (1889–1977)
—George Lemaître (1894–1966)

Part of growing-up is coming to terms with your place in the world. For science, growing-up has meant discovering where we are in the universe. At the beginning of the 20th century there were two schools of thought. One claimed that Earth existed within the only galaxy; the other camp believed that there were many galaxies. Through careful observations, Edwin Hubble proved that there are numerous galaxies spread throughout the universe. What was more startling was that he showed that the galaxies are moving away from each other, as the universe itself expands. The findings shocked Einstein and brought Hubble international fame.

From law to big lenses

Some people succeed in everything they turn their hand to. Hubble was one of these. Having been born in Missouri and them moved to Chicago, he didn't work particularly hard at school, but still got a scholarship to study at the University of Chicago, despite his head teacher's comment that he had "never seen [him] study for 10 minutes." He must have worked a little harder at athletics, because he broke the Illinois state high-jump record.

Once at university he didn't allow his academic work to get in the way of sport, and played in the basketball team. Nevertheless, again he excelled, and was given a Rhodes scholarship, allowing him to travel and study. He chose Oxford as his destination and law as the subject, having promised his dying father to train as a lawyer. Again he fitted sport into his curriculum, at one point having a boxing match with French fighter Georges Carpentier. Arriving back in America just before the outbreak of World War I he took a job as an attorney, but was just about to switch to astronomy when called to fight in Europe.

So it was only after traveling around much of the world that Hubble started working at the world's leading observatory, where he would discover that the universe was bigger than anyone had previously thought.

He arrived at the Mount Wilson Observatory near Pasadena, California, just three weeks after they had unveiled their new telescope. The Hooker telescope was the biggest in the world and collected light using a massive 100-inch mirror. It had taken 10 years to plan and build. The telescope was in high demand, and each member of staff had to take their turn to perch on the small viewing platform high above the concrete floor beneath the open observatory dome.

The universe is bigger

It's easy to live on planet Earth and forget how it fits into the bigger picture. The Earth orbits a smallish star, the Sun, along with eight other planets. The Sun, however, is only one of some one hundred billion stars that make up our galaxy. If we could travel away from our galaxy, we would see that the stars are arranged in spiraling arcs, much like the sparks flying from a Catherine wheel firework. Observing it from Earth, we can only look sideways through the disk and the stars appear as a bright band across the sky — the Milky Way.

Hubble started taking photographs of one particular cloud-like patch of light that appears to be about the size of the Moon. Called the Andromeda nebula (*nebula* is Latin for cloud), no one was sure what this mass was, though many thought that it was a cloud of luminous gas situated within the galaxy.

The assumption was reasonable enough, as Harlow Shapely, a former worker at Mount Wilson, had previously provided evidence indicating that the galaxy in which our solar system exists was the only galaxy. In addition, Shapely had estimated the size of the galaxy as being an astonishing 300,000 light-years in diameter. Although this is vast, telescopes that existed before the Hooker would have been able to reveal individual stars within Andromeda if it were a cluster of stars, so it must be a ball of gas.

In October 1923, Hubble spotted several bright points of light flaring dramatically within the Andromeda nebula. They looked remarkably like novae, stars that eject part of their material and dramatically increase their luminosity. Clearly there were at least some objects in this cloud. He went to the records of previous photographs taken of Andromeda and studied them carefully. By studying old photos, and taking new ones whenever he had the opportunity, Hubble found that one of the points of light flared on a regular cycle.

This was a critical breakthrough. These regularly cycling stars, Cepheid variables, have an intriguing property. They exist in two states. In one, the star is compact and at high temperature. Pressure builds up and causes it to expand. As its size increases, so does the amount of light it gives out. This state is unstable and the star collapses back to its compact state again.

By studying these stars, astronomers are convinced that the amount of light they give out is directly proportional to the frequency of the cycle. Cepheid variables that go from bright to dim every other day are not very bright, while those that take 100 days to perform each cycle are extremely bright. This makes the stars very useful, because if you measure how much light is reaching the Earth and then compare it with the amount you calculate left the star in the first place, you can work out the distance to the star.

Using this calculation, Hubble became convinced that the Andromeda nebula was one hundred thousand times further away from us than the most distant star in our own galaxy. The implication was colossal. The universe contained at least two galaxies — possibly more.

Classification

Hubble wasn't the first person to suggest that nebulae were separate galaxies, but he was the first to produce proof. Before him various people had studied these clouds of light and started to develop systems of classification.

The most extensive survey of the sky before the invention of photography had been carried out in the 18th and 19th centuries by Caroline Lucretia Herschel, her brother Frederick William Herschel, and his son John Frederick William Herschel. Together they discovered some 4,630 star clusters and nebulae, bringing the total number of known clusters to 5,079 — a collection that John Herschel published in *Philosophical Transactions* in 1864.

In 1908, the German astronomer Maximilian Franz Joseph Cornelius Wolf had started to classify these clusters using photographs taken by a telescope at Heidelberg. He developed a system in which clusters that had little form had letters at the beginning of the alphabet, while those with clear spiral patterns within their shape had higher letters.

It was a start, but Hubble was unimpressed. In his Ph.D. thesis he commented that the system "offers an excellent scheme for temporary filing until a significant system shall be constructed." He felt that it was an empirical system that had no physical significance and that it was only a matter of time before he did a better job. The solution that he announced in the years between 1922 and 1926 divides galaxies into elliptical, spiral, barred spiral, and irregular. These are subdivided into categories a, b, and c according to the size of the central mass of stars within the galaxy and the tightness of any spiraling arms.

Velocity, distance, and expanding universe

In itself, this would have been enough to place Hubble in the textbooks, but he had more. Hubble began to measure the distance between the Earth and the galaxies he identified. Several astronomers, in particular American Vesto Slipher, had been studying the speed at which nebulae were moving relative to the Earth for many years before Hubble had given proof that they were galaxies.

The method they used was to look at the color of the light that was coming from the galaxies. They had noticed

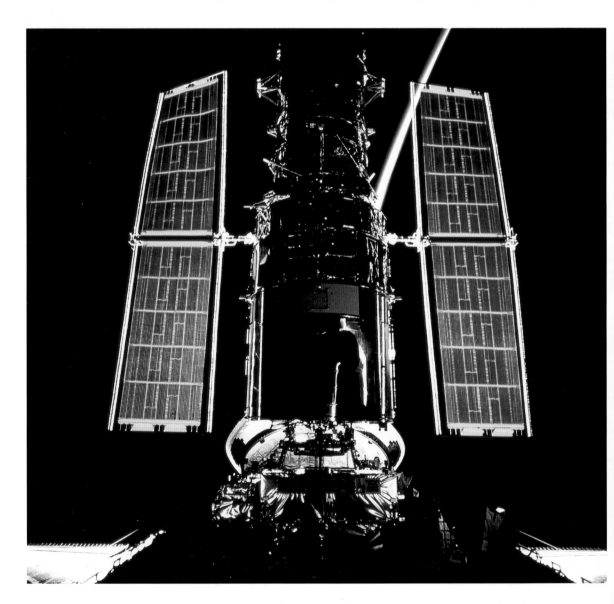

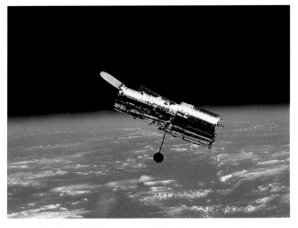

above *The Hubble telescope in orbit — the most sensitive optical telescope yet constructed.*

opposite *Named in his honor, the Hubble Telescope was carried into orbit by the space shuttle Discovery on April 25, 1990. After needing modification due to manufacturing errors, it has provided amazing views of the universe.*

Timeline

1889
Born on November 20 in Marshfield, Missouri, USA
1898
Family moves to Chicago
1906
Receives a scholarship to the University of Chicago to study mathematics and astronomy
1910
Receives his B.Sc. degree and is awarded a Rhodes scholarship, which he uses to study Roman and English Law at Queen's College, Oxford
1913
Returns to the United States and becomes an attorney in Kentucky
1914
Turns down a position at the Mount Wilson Observatory, Pasadena, California, to join the army and serve in World War I. He becomes a major in the 343rd Infantry 86th Division and serves in France and Germany
1919
Returns to the United States and joins the staff of the Mount Wilson Observatory, Pasadena, California
1920
Publishes his Ph.D.
1922
Presents an outline concept of a new way of classifying galaxies
1924
Marries Grace Burke, and in the same year announces his discovery of a Cepheid variable, in the Andromeda nebula, which allows him to calculate that the nebula is 100,000 times further away than the nearest stars
1929
Discovers the "red-shift" and shows that the universe is expanding
1942
Leaves the observatory to do war work at the Aberdeen Proving Ground
1946
Is awarded a Medal of Merit and returns to the Mount Wilson and Palomar Observatories
1953
Dies on September 28 in San Marino, California

that the light was fractionally more red than they predicted it should have been — a feature they termed "red-shift." They attributed this change in color to the Doppler effect. This is easily experienced with sound waves, where, if you stand beside a road and listen to an approaching car, the pitch of the noise seems higher when it is coming toward you than after it has passed and is going away. In effect the movement of the car means that the sound waves reach you at increased frequency if the car is approaching and at decreased frequency when it departs.

The same process occurs for light, but the object has to be traveling at an enormous speed for this to be noticed. This is exactly the case with the galaxies, which are moving rapidly away from each other. Consequently, light reaching the Earth appears to be of a lower frequency, and as such it appears to be a little redder. The greater the change, the greater the speed at which the Earth and the observed galaxy are separating.

Into the melting pot of ideas came Einstein's theory of general relativity. His idea that space is curved by gravity predicted that the universe — the area containing the total collection of galaxies — could either be expanding or contracting. It could not be static. It was an idea so preposterous that even Einstein refused to believe it was possible, and modified his original theory to try to avoid the problem.

By studying the red-shifts, Hubble and his assistant Milton Humason showed that Einstein should have had more confidence in his work. The universe was indeed expanding, and an elated Einstein freely admitted that his attempt to bend his theory had been the greatest blunder of his professional life.

Hubble also realized that there was a precise correlation between the position of a galaxy and the speed of its movement. He found that the galaxies that were in the outermost areas of the universe were moving with greater speed than those nearer the middle. This feature is now referred to as Hubble's law.

Not only was the universe expanding: the public recognition of Hubble's work meant that he and his wife found that their social circle had become very wide. Living on the edge of Hollywood, with his own version of celebrity status, led him to form friendships with the likes of Charlie Chaplain, Helen Hayes, and William Randolph Hearst, as well as luminaries such as the author of *Gentlemen Prefer Blondes*, Anita Loos. He had studied the stars and in return had become a star himself.

George Gamow

(1904-1968)

Acquaintances
- Neils Bohr (1885–1962)
- John Cockcroft (1897–1967)
- Friedrich Georg Houtermans (1903–1966)
- Ernst Thomas Sinton Walton (1903–1995)
- Robert Atkinson (1929–1993)
- Edward Teller (1908–)
- Ralph Alpher (1921–)

If Edwin Hubble was right, and the universe was expanding, there must be a point in history when this expansion began. There must also be a point in the universe that was the focus of this beginning: presumably any event large enough to generate a universe containing many thousand of galaxies would be huge. To call it a Big Bang is something of an understatement, but the term has popular use, and George Gamow, together with his student Ralph Alpher, sorted out a mathematical understanding of how it could have occurred. His breakthrough started when he tried to understand alpha particles.

It's all about protons

At the end of the 1920s, scientists were puzzled about alpha particles. The particles had the ability to escape from radioactive nuclei, which, according to currently held theories, should not have been possible.

Physicists were convinced that alpha particles were held in the nucleus by "strong forces." As the name suggests, these forces are powerful, but they can only operate over short distances. If a particle gets even fractionally outside the nucleus it would be outside the reach of these forces and would be free to fly away. Also, because the particle is positively charged it will be actively repelled by the positively charged nucleus. But how could it make that first move? In order to be able to think about the issue, physicists considered the situation to be as if the alpha particle was a solid ball, sitting in a crater on the top of a volcano. If you push the ball up to the crater's rim and let it go, it will roll away. For years they tried to find a way of explaining where the energy needed to "lift" the particle came from.

Then came Gamow. He took a different approach and invoked quantum physics. This allowed him to treat each particle also as a wave. Because waves are spread out, their location is not restricted to any one point. A wave could therefore spread right through the crater walls, effectively allowing the particle to tunnel out and escape.

The idea may seem fanciful, but in the world described by quantum physics it works. In addition, Gamow found that this theory explained how, within the Sun, the waves of two hydrogen atoms (protons) can combine to form a single alpha particle, or helium nucleus, in effect performing the reverse process. Furthermore, this nuclear reaction would generate heat, explaining the source of the Sun's heat.

Big Bang

In 1927, the Belgian priest George Lemaître had proposed that the universe began about 10–15 billion years ago with an explosion of a "primeval atom," now more often referred to as a "singularity." The key ingredients involved in this event were believed to be protons, neutrons, and electrons. Gamow began to wonder how protons would have behaved under the extremes of heat and pressure that must have occurred in the super-dense fireball of this Big Bang.

By the 1940s Hubble had shown his evidence of an expanding universe, but few people were prepared to put their trust in the observations or his subsequent calculations. Scientists love to be able to repeat an experiment, and for obvious reasons no one was about to redo the Big Bang.

The only option was to start with what existed at the moment and see if it was possible to find a plausible explanation of its origins. One aspect of the observable data was that stars are made of 75% hydrogen and 25% helium. Another is that these two gases make up over 99% of all the visible material in the universe. Gamow found it easy to

Time after The Big Bang

10^{-35} Sec	THE BIG BANG
10^{-6} Sec	Universe Shaped
3 Sec	Basic Elements Form
10,000 Yrs	Radiation Era
300,000 Yrs	Matter Domination Era
300 Million	Stars and Galaxies Form

Year before the present

5 Billion	Birth of the Sun
3.8 Billion	Earliest Life Forms
700 Million	Primitive Animals Appear
200 Million	Mammals Evolve
65 Million	Dinosaurs are Extinct
600,000	Homo Sapiens Evolve
170,000	Supernova 1987A Explodes

AD

1054	Crab Supernova Appears
1609	Galileo Builds First Telescope
1665	Newton Describes Gravity
1905	Einstein Publishes Relativity
1929	Hubble Discovers Universe Expansion
1960	Quasars Discovered
1964	Microwave Radiation Discovered
1967	Pulsars Discovered
1987	Supernova 1987A Reaches Earth
1990	Hubble Telescope Launched
1990	Big Bang Confirmed
1993	Hubble Telescope Repaired

Years in the future

100 Trillion	Stellar Era Ends
10^{37}	Degeneration Era Ends
10^{38} to 10^{100}	Black Hole Era Begins
10^{100}	Dark Era Begins

above *Stars are made up of hydrogen and helium — two gases that make up almost all of the visible material in the universe.*

explain how hydrogen could form, as this needs only protons and electrons to come together. The problem was in describing the origin of the helium.

Working with his assistant Ralph Alpher, Gamow found a way of describing the conditions that would lead to a universe containing the existing proportions of hydrogen and helium. He also showed that as the universe expanded, the density would drop rapidly and the nuclear reactions would halt. This meant that more complex elements could not be formed. Although this was clearly a problem, Gamow was content because, as he said, the equations dealt with 99 percent of the known matter in the universe.

Amazingly, the equations contained only fundamental parameters, such as Newton's gravitational constant, Planck's constant, the charge and entropy quanta, and the binding energy of deuteron (a form of hydrogen). It took Fred Hoyle a few years later to sort out where the one percent of complex elements came from.

Their calculations showed that as long as you set the temperature precisely it was quite possible to match the proportion of hydrogen and helium seen in the universe. The equations also show how the temperature of the "Bang" will change with time. Using the kelvin temperature scale, the current form of this equation (which has been slightly modified since its arrival in 1948) sets the temperature as 10 billion divided by the square root of the age of the universe in seconds. This means that one second after the "Bang" the temperature would have been 10 billion degrees. After 100 seconds it would already have cooled to 1 billion degrees, and an hour later it would be down to a mere 170 million degrees. Gamow established the currently accepted view that the temperature of the "Bang" was about 5 degrees kelvin — that's −268°C.

Cold leftovers

In 1964 the German-born American astrophysicist Arno Allan Penzias and colleague American physicist Robert Wilson made a discovery that thrilled Gamow. Using a large radio telescope that was designed primarily to work with communication satellites, they found that wherever they looked in the universe there was a gentle "hum." The frequency of it corresponded to the radiation that would

be expected from a black-body that was at 3.5° kelvin. This was the background radiation that Gamow had predicted, and the discovery won Penzias and Wilson the 1978 Nobel Prize for Physics. Nobel Prizes are never awarded posthumously, so the citation didn't include Gamow's name.

In November 1989, NASA launched a satellite called the Cosmic Background Explorer (COBE). Its purpose was to orbit the Earth and take a detailed look at the radiation in the universe. In 1992, COBE discovered "ripples" in this radiation, another feature that Gamow had predicted.

Although the Big Bang has not been proved, and probably never can be, the evidence is massing, and Gamow played a pivotal role in discovering it.

Timeline

1904
Born in Odessa, Russia, on March 4
1922
Studies at Novorossia University, Odessa
1923
Studies at the University of Leningrad, where during the course of his Ph.D. he looked to see if the newly formulated quantum theory could also be applied to the nucleus of atoms
1928
Takes a fellowship in theoretical physics at the University of Copenhagen and proposes that atomic nuclei can be treated as little droplets of "nuclear fluid." These views led ultimately to the present theories of nuclear fission and fusion
1929
Moves to Cambridge University as a Rockefeller Fellow
1930
Returns to Copenhagen, this time as a fellow of Neils Bohr's Theoretical Physics Institute
1931
Marries Lyubov Vokhiminzena and becomes a professor at the University of Leningrad
1933
Becomes a fellow of the Pierre Curie Institute, Paris, and a visiting professor at the University of London
1934
Moves to the United States and takes a professorship at the George Washington University in Washington D.C. where he develops the theory of the internal structure of red giant stars
1935
His son, Rustem Igor, is born
1942
Works with Teller on a theory of the internal structures of red giant stars
1956
Moves to the University of Colorado and is divorced from Lyubov
1958
Marries Barbara Perkins
1968
Dies on August 19 in Boulder, Colorado

Fred Hoyle

(1915–2001)

Acquaintances
— William Alfred Fowler (1911–1995)
— Raymond Alfred Lyttleton (1911–1995)
— Hermann Bondi (1919–)
— Thomas Gold (1920–)
— Margaret Burbidge (1922–)
— Geoffrey Burbidge (1925–)

The term the "Big Bang" was probably coined by Fred Hoyle, though, paradoxically, Hoyle didn't believe in the theory. Instead, he maintained that the universe has existed for an infinite past and would continue infinitely into the future. He held this controversial Steady State theory despite much criticism. It wasn't his only point of controversy: he was convinced that the seeds of life had been created elsewhere in the universe and traveled here, possibly on a meteor. His most valuable contribution to science, however, was an understanding of how elements could form within stars.

From infinity to infinity

During World War II, Fred Hoyle found himself working in the Admiralty Signals Establishment alongside Hermann Bondi and Thomas Gold. Together they considered theories about how the universe had come into existence. They were unimpressed with the concept that the universe had started with some colossal explosion, and that before that there had been nothing at all. It seemed too far-fetched to be true, and at that point there was very little, if any, hard evidence.

Instead, they developed what they called the continuous creation theory. In 1948 Bondi and Gold published a paper that expounded the idea, and two months later Hoyle published another. While Bondi and Gold's paper concentrated on the philosophical aspects of a universe that had a high degree of uniformity in terms of time as

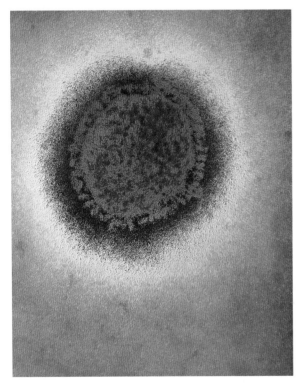

above *The influenza virus consists of a core of ribonucleic acid (red) surrounded by a spiked protein envelope of haemagglutinin (brown). The spikes can change structure to create a new influenza strain such as this Beijing strain which spread as an epidemic in 1993.*

well as space, Hoyle's dealt with the creation of hydrogen. He believed that to maintain a steady state you would need a continual supply of primordial hydrogen and provided evidence from the theory of general relativity to show how this might be possible.

At first the theory was held as a possibility alongside the Big Bang concept, but like many scientific theories it ran into a problem—the admission of new data. The 1965 discovery of background microwave radiation throughout the universe added such weight to the Big Bang, that it became very difficult to keep believing in the Steady State model. Difficult, but not impossible, and Hoyle never accepted defeat, constantly looking for weaknesses in the Big Bang theory.

Creating elements

While Hoyle's Steady State concept got considerable public attention, his most important contribution to science was in the ideas he generated about the way that elements could have been created. Gamow's concept of a massive rush from a singularity, a tiny zone of incredibly hot, high-pressure material, provided an understanding of how hydrogen and helium could have come into being. But this left out all of the other more complex elements that are so vital in forming life and other systems.

The problem for Gamow was that the universe expanded and cooled down too fast. Working at the Calfornia Institute of Technology in Pasadena with W.A. Fowler and with husband and wife team Geoffrey and Margaret Burbidge, Hoyle came up with an explanation that required only the temperature found in the interior of stars. The theory accounted for the different distribution of elements found in the universe and provided an explanation of the way that stars evolve.

This cooler theory of nucleo-synthesis had the distinct advantage for Hoyle that it would operate within his concept of a Steady State universe.

Shunned for belief in ET life

The work on the creation of elements was published as a paper in 1957, listing as the authors the two Burbidges, Fowler, and Hoyle. It was widely recognized that the intellectual content of the paper had come primarily from the latter two, so there was an element of surprise when the Nobel Prize for this work was given only to Fowler.

The only plausible explanation for this was offered by the editor of the science journal *Nature*, who suggested that the Nobel committee did not want to give credit to Hoyle

Fred Hoyle, Geoffrey Burbidge
and Jayant V. Narlikar

A Different Approach to
Cosmology
from a Static Universe through
the Big Bang towards Reality

above *With its subtitle of "From a Static Universe through the*
Big Bang towards Reality," Hoyle and his two co-authors
challenge many of the accepted views of cosmology, and
question whether the Big Bang ever occurred.

because of other ideas that he held. Hoyle believed passionately that life must have sprung up all over the universe. He argued that the initial molecules that enabled living things to evolve on earth had come originally from elsewhere in the universe, traveling here through space. He was convinced that the universe contained other intelligent beings. During a radio broadcast in the early 1950s, at a time when Australia was dominating England at cricket, he even remarked that he would wager that somewhere in the Milky Way there was a cricket team who could beat the Australians.

This alone was not enough to upset the authorities, but his theory about influenza epidemics caused great concern. Massive pandemics of influenza occur at intervals of between five and 50 years. When a pandemic strikes, the result can be devastating. The 1918 pandemic of so-called Spanish 'flu killed an estimated 20 million people. Hoyle maintained that large epidemics of influenza occurred when planet Earth passed through certain meteor streams, and viruses never encountered by human immune systems came to Earth on board meteor dust.

The concept caused widespread fear, but criticisms were easy to frame, because Hoyle had also started writing popular science fiction, and the scientific authorities claimed that he was confusing his fact and his fiction.

Even so, he was highly respected, and when he was awarded the Royal Medal by the Royal Society the then president complimented the originality evident in all of his work, saying that his "enormous output of ideas are immediately recognized as challenging to astronomers generally... his popularization of astronomical science can be warmly commended for the descriptive style used and the feeling of enthusiasm about his subject which they succeed in conveying." He certainly enriched science, even if he did goad a few scientists.

Timeline

1915
Born in Yorkshire on June 24. He was the son of a wool merchant, and by the age of 10 he could navigate by the stars. After attending his local grammar school he gained a scholarship enabling him to go to Emmanuel College, Cambridge, to read mathematics

1939
Is elected to a fellowship at St. John's, Cambridge, and marries Barbara Clark. The marriage was to produce a son, Geoffrey, and a daughter, Elizabeth

1940
During the war he works with Hermann Bondi and Thomas Gold for the Admiralty on projects such as radar

1945
Returns to Cambridge as a lecturer in mathematics

1957
He is elected a fellow of the Royal Society

1958
He is appointed the Plumian Professor of astronomy

1963
Takes an appointment as a visiting professor at the California Institute of Technology

1967
Becomes the first director of the Cambridge Institute of Theoretical Astronomy

1972
Hoyle resigns his professorship in a dispute over the election of another professor
Receives a knighthood

1973
He resigns from the Institute of Theoretical Astronomy and moves to the Lake District, before moving to the south coast of England

1974
He is awarded the Royal Medal of the Royal Society

2001
Dies on August 20 in Bournemouth, England

Place an ice cube on a table. It melts and forms a pool that drips onto the floor. Nothing particularly surprising. But you would be surprised if the reverse occurred, if the pool on the floor rose to the table and formed a frozen cube. Time, says Ilya Prigogine, moves in one direction, as if it is an arrow that points from the past to the future. To an extent this seems obvious, but when introduced, the concept broke the standard understanding of thermodynamics. We are very used to watching these processes occur, but Prigogine maintains that the arrow of time is driven by the increase of entropy.

Ilya Prigogine

1917–

Acquaintances
— Théophile De Donder (1873–1957)
— Jean Timmermans (1882–1971)
— Alan Mathison Turing (1912–1954)

Time's arrow

When, in 1946, Russian-born Ilya Prigogine challenged physics by suggesting that its obsession with equilibrium was the cause of a critical misunderstanding, he was roundly condemned by the establishment. Prigogine often recalls one senior scientist who stood up after a lecture and said, "Why is this young man devoting his interests to irreversible causes? Irreversible causes are just illusory. Time is just a parameter; so forget about it." But Prigogine was convinced—time wasn't just a parameter in an equation that could be played forward or backward. Time was irreversible and that had consequences.

The chance of order arising from disorder is infinitesimally small, according to classical physics. The standard argument is that it goes against the well-established laws of thermodynamics, which predict that if left alone, everything tends towards a state of uniform disorder. Applying thermodynamics to the universe suggests that the probability that an explosive rush of energy, the Big Bang, could generate an ordered system of galaxies, stars, planets, and, on at least one planet, of living systems, is so small that it could not occur. Prigogine, however, provided a solution to the problem, saying that the clue lies in the fact that the universe is a "far from equilibrium system."

Thermodynamics made sense at the level of steam engines. Burning a certain quantity of fuel releases a specific amount of heat. Some of this heat is employed for work, and the rest dissipates into the environment. As heat is the effect of atoms moving at random, increasing the general level of heat in the environment simply increases the amount of random activity, a process that physicists call an increase in entropy. This increasing entropy is Prigogine's arrow of time.

Prigogine's work indicates that there are two types of processes. Some, such as planetary motion, are time-reversible. They obey Newton's laws and you can run equations forward or backward to show where planets have been in the past, or where they will be in the future. Other systems, like heat conduction or chemical reactions, do not show this symmetry, they are non-equilibrium systems. Prigogine has shown that these irreversible processes play a fundamental role in our universe and can generate coherent behavior and structure.

The problem of order

The issue, said Prigogine, becomes clear if you separate micro and macro systems. In the case of a steam engine, entropy increases and energy is conserved if you consider the whole process—the macro system appears to be in equilibrium. But look at any of the individual components, and each micro system will be far from equilibrium.

Another prime example of non-equilibrium is provided by living organisms. They consume energy in the form of nutrients, perform work, and excrete waste. Without being changed they can give off heat to their surroundings. He was fascinated by the way embryos develop—a clear case of a system creating order.

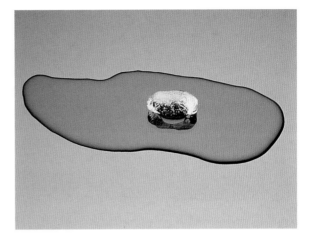

above *We are used to seeing ice melt into a puddle, but we would be stunned if a puddle turned into a cube of ice.*

In his Nobel Prize autobiography Prigogine wrote: "It appeared to me that living things provided us with striking examples of systems that were highly organized and where irreversible phenomena played an essential role."

Evidence supporting his theories came from the work of two Russian scientists, Boris P. Belousov and Anatol M. Zhabotinsky, who discovered an interesting chemical reaction. When they mixed malonic acid with an acidified bromate solution and a metal ion catalyst the mixture oscillated between a colored and a clear state, spending about one minute in each state. The reaction required a continuous source of energy, but it behaved as a self-organizing system even though it was inanimate. When held far from equilibrium, the system was quite capable of creating order from chaos.

An amusing application of Prigogine's idea is the theory of traffic flow. With a small number of cars on the road, everyone can drive as they want. Viewed from above, the traffic flow will be disordered. Increase the density and it all changes. There will be a critical point where the drivers are forced to drive in a uniform manner, thereby losing their independence. Viewing this system from above, it will now appear much more organized—order has appeared out of chaos. This is another example of a "dissipative structure," a structure that exists only in conjunction with its environment.

The Nobel committee summed up Prigogine's contribution to science, saying, "[He] has fundamentally transformed and revised the science of irreversible thermodynamics. He has given it new relevance and created theories to bridge the gap that exists between biological and social scientific fields of inquiry."

In the last years Prigogine has developed what he calls microscopic laws that include the arrow of time. His desire is to extend the fundamental laws of physics so that they include the arrow of time, and in so doing make sense of the observations that come from areas of science as distant as fundamental particles and cosmology.

For centuries philosophers have studied and argued about the fundamental mechanisms of orderly precise reasoning, the process of logic. Self-taught English mathematician George Boole, however, said that logic should be a branch of mathematics rather than philosophy. His so-called Boolean algebra uses mathematical symbols rather than words to express logical relations. The idea was revolutionary and won him a professorship in Cork, Ireland. Boolean algebra now forms the basis of electronic circuits and telephone switching equipment, as well as providing the foundation for the library classification of books and Internet search systems.

George Boole

1815–1864

Acquaintances
—Augustus de Morgan (1806–1871)
—Mary Boole, née Everest (1832–1916)
—Charles Sanders Peirce (1839–1914)

From metaphysics to mathematics

Finding ways of developing a logical argument had been a critical part of philosophy ever since the days of Plato and Aristotle, but the methods employed complex wordy arguments. George Boole, however, grew up with little knowledge of previously determined systems of logic and debate. As the son of an unsuccessful small shopkeeper, he had little formal education and ended up teaching himself mathematics. While this made the early years of his thinking hard, it gave him the advantage of being relatively unrestricted by previous ideas.

Consequently, Boole argued that there was a close analogy between algebraic symbols and symbols that represent logical interactions. He also showed that you could separate symbols of quality from those of operation. He cut through all the complexity and developed a "process of analysis" that allowed people to break thought processes into individual small steps. Each step involves making some proposition that is either true or false. True is given a value of 1 and false 0. The answers from these steps can be combined by using one of three operators: AND, OR, or NOT.

At its simplest, the idea is that you take two proposals at a time and link them with an operator. The AND operator gives a value of 1 only if both of the original proposals had produced answers of 1. If they gave a 0 and a 1 then the answer from AND would be 0. The OR operator gives an answer of 1 if either original proposal was 1, and the NOT operator gives a value of 1 if neither original proposal had been attributed 1. By adding many steps, Boolean algebra can form complex decision trees that produce logical outcomes from a series of previously unrelated inputs.

Publication, praise, and pneumonia

While working as a teacher in various village schools in Yorkshire, Boole started writing papers on aspects of mathematics. With no formal training he found it difficult to get them accepted by publishers. One editor, Duncan Gregory, who edited the recently founded *Cambridge Mathematical Journal*, was impressed by his work and encouraged him to take some courses at Cambridge. Unfortunately, Boole hadn't the finances to make this possible, but it was Gregory who stimulated and encouraged his interest in algebra.

Soon Boole was producing work that equaled that of the best academics, and in 1844 he was awarded the Royal Medal of the Royal Society. This was the first time that the medal was awarded for pure mathematics.

In 1847, Boole published his first book, *Mathematical Analysis of Logic*. It was a small volume but presented his idea that logic was better handled by mathematics than metaphysics. At this point he was teaching in a school in Lincoln, but the book brought his work academic acclaim. Without having any degree he was offered the post of professor of mathematics at Queen's College, Cork. He used the security this bought to develop his ideas and publish his masterpiece: *An Investigation of the Laws of Thought*.

Boole's life sadly came to an abrupt end when he arrived home one evening after giving a lecture. Having walked two miles through drenching rain, he caught pneumonia. Unfortunately, his wife had a belief that the best way of treating an illness was to expose the patient to the original cause. She therefore repeatedly poured water over his bedclothes to ensure they remained wet. Unsurprisingly, he failed to recover.

Before his death, Boole had published around fifty papers and had begun to investigate the basic properties of numbers that underlie algebra. Even though he was clearly excited by his work, without a doubt he could never have foreseen the explosion in use of his ideas that occurred with the advent of computers — which primarily operate by asking yes/no questions, and represent the answers in a series of 1s and 0s.

Timeline

1815
Born on November 2 in Lincoln, England

1835
Starts to teach himself mathematics, studying the works of Isaac Newton (1642–1727) as well as French mathematicians Pierre Simon Laplace (1749–1827) and Joseph Louis de Lagrange (1736–1813)

1849
Is appointed professor of mathematics at Queen's College, Cork, Ireland

1854
Publishes *An Investigation into the Laws of Thought, on Which are Founded the Mathematical Theories of Logic and Probabilities*

1855
Marries Mary Everest, niece of Sir George Everest, after whom the mountain is named. Over the following nine years they have five daughters — Mary, Margaret, Alicia, Lucy, and Ethel

1857
Is elected a fellow of the Royal Society

1859
Publishes *Treatise on Differential Equations*

1860
Publishes *Treatise on the Calculus of Finite Differences*

1864
Dies of pneumonia aged 49, on December 8 in Ballintemple, County Cork, Ireland

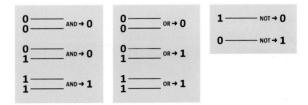

above *Boolean logic allows two or more results to be combined into a single outcome. This lies at the center of microelectronics as well as Internet-based search engines.*

Mathematics, said David Hilbert, is concerned with formal symbolic systems. It is an activity that uses a series of symbols and rearranges them according to various formal rules. This so-called formalism is an extreme view of mathematics that separates it from any concrete reality. Consequently there is nothing external to its workings that can be used to validate it, so all of its arguments must be capable of justifying themselves. Speaking in Paris in 1900 Hilbert caused a stir by presenting 23 mathematical problems, some of which are yet to be solved more than a hundred years later.

David Hilbert

1862–1943

Acquaintances
— Paul Albert Gordan (1837–1912)
— Felix Christian Klein (1849–1925)
— Carl Louis Ferdinand von Lindermann (1852–1939)
— Adolf Hurwitz (1859–1919)
— Hermann Minkowski (1864–1909)
— Albert Einstein (1879–1955)
— Norbert Weiner (1894–1964)

A radical approach to mathematics

Russian-born David Hilbert has gone down in history as the person who introduced the concept of formalism into mathematics. As a mathematician, philosopher, and physicist, he reduced branches of mathematics such as geometry to a series of axioms, of basic self-justifying principles. For Hilbert, the beauty of this system of thought was that it removed mathematics from any relationship to a physical reality. It was all about symbols on paper and rules for manipulating those symbols.

A criticism of formalism is, however, that it denies that mathematics can shed any light on natural science. This goes against clear observations that mathematics can be used to predict natural events. Despite this, the pursuit of formalistic ideas led to many developments within mathematics.

Some things don't change

Part of Hilbert's contribution to mathematics is that he introduced an original approach to ways of considering mathematical invariants. An invariant is something that is left unchanged by some class of functions. In terms of a geometrical transformation, an invariant would be an object that does not alter its shape or size while it is being moved. His methods allowed the structure of mathematical theories to become the subject of mathematical analysis. He proved that all invariants could be expressed in terms of a finite number — a number that can actually be counted.

He spent the first two decades of the 20th century struggling to construct a self-justifying system of arguments that would prove that a finite number of steps of reasoning could not lead to a contradiction.

Sadly for Hilbert, this work was itself contradicted in 1931 when Czech-born Kurt Gödel published his Incompleteness Theorem, showing that every consistent theory must contain propositions that are undecidable. Gödel pointed out that when proving statements about a mathematical system at least some of the rules and axioms must derive from outside that system. But by doing this you create a larger system that will contain its own unprovable statements.

The implication is that all logical systems of any complexity are, by definition, incomplete. Some people have extended this argument to claim that it will be impossible to create a computer that is as capable as a human being, because the computer will always be limited by a fixed set of axioms, whereas people can discover unexpected information or truths. In doing this, Gödel showed the intriguing principle that truth is more important than provability.

Problems are the life-blood of mathematics

Much of Hilbert's fame rests on a list of 23 problems that he presented at the Second International Congress of Mathematicians in Paris at the dawn of the 20th century. Many of them have now been solved, and each time this happens there is a notable level of excitement within academic mathematics.

He saw the problems as opportunities and signs that mathematics was a living subject with a great future. "The great importance of definite problems for the progress of mathematical science in general ... is undeniable ...[for] as long as a branch of knowledge supplies a surplus of such problems, it maintains its vitality ... every mathematician certainly shares the conviction that every mathematical problem is necessarily capable of strict resolution ... we hear within ourselves the constant cry: There is the problem, seek the solution. You can find it through pure thought."

There has been some debate over whether Hilbert or Albert Einstein discovered the correct equations for general relativity first. In fact historians are now confident that Einstein is correctly attributed with winning the race, but there is every good reason to believe that Hilbert was not far behind.

In 1930, when he retired from the University of Königsberg, Hilbert gave a speech which ended with six famous words: "Wir müssen wissen, wir werden wissen" — "We must know, we shall know."

Timeline

1862
Born in Königsberg, Prussia (now Kaliningrad, Russia) on January 23

1880–1885
Works for his Ph.D. at the University of Könisgsberg

1886–1895
Works as a member of staff at the University of Könisgsberg and is given a professorship in 1893

1888
Works on invariant theory and proves his famous basis theorem

1893
Begins work on algebraic number theory

1895
Moves to Göttingen, Germany, taking the highly prestigious post of professor of mathematics

1897
Publishes *Zahlbericht*, a synthesis of the work of Kummer, Kronecker and Dedekind

1899
Publishes *Grundlagen der Geometrie*, putting geometry on a formal axiomatic setting

1900
Presents 23 problems in his speech in Paris entitled "The Problems of Mathematics." Some have still not been solved

1930
Hilbert retires, and the city of Königsberg makes him an honorary citizen

1943
Dies on February 14 in Göttingen

Biology and mathematics are often seen as two very different pursuits. Norbert Weiner, however, brought the two together with his work in communication theory. This led him to study the systems of control and communication that operate in animals and machines. He introduced the concept that systems have inputs and outputs that are influenced by feedback mechanisms—an understanding that has had applications in a multitude of different applications. His work also formed the basis for the branch of artificial intelligence based on the processes of the human mind. As such, he invented the discipline of cybernetics.

Norbert Weiner

1894–1964

Acquaintances
— David Hilbert (1862–1943)
— Walter Cannon (1871–1945)
— Bertrand Russell (1872–1970)
— Vannevaar Bush (1890–1974)

A fast-track genius

Having been born in Columbia, Missouri, on November 26, 1894, Norbert Weiner made a flying start in academic life. This is just as well, as his father, a somewhat eccentric professor of Slavic languages and literature, was determined to raise a genius. Aged nine he entered high school and completed the four-year course in two years. He had completed his undergraduate degree at Tufts University by the age of 14, at 18 he got his Master's degree and a year later completed his Ph.D. in mathematical logic.

With this impressive track record he traveled to Cambridge University to study logic with the controversial Welsh philosopher, mathematician, and prolific author Bertrand Russell, a recognized world-leader in the subject, before spending a few months in Germany with mathematician David Hilbert. Returning to the U.S., Weiner started to find it difficult to concentrate on academic work and his mind wandered. He taught at Columbia, Harvard, and Maine Universities, before moving to work as a staff writer at the Encyclopedia America and then spent an unhappy year at the Boston Herald.

At that point World War I broke into his life. As an ardent pacifist he took a post outside military ranks at the U.S. army's Ballistic Research Laboratory in Aberdeen, Maryland. The experience he gained there, formulating mathematical tables that determine where guns should be aimed before firing, rekindled his interest in mathematics and in particular in theories of control.

When in 1919 he was appointed a professor at Massachusetts Institute of Technology, he started researching the way that moving particles interact with each other. It might seem a limited area of research, but for Weiner it was research into communication. If you look at the way things interact, you see how this interaction affects each item. In effect you are monitoring the way that they communicate. Once you have built up a framework for investigating systems you can apply it to animate or inanimate systems—to animals or machines. He called his new-found discipline cybernetics, deriving it from the Greek word for steersman, *kybernetes*.

The importance of feedback

When World War II came along, Weiner found himself developing new systems for controlling anti-aircraft guns. He developed ways for guns to draw information about a target's speed and direction, and combine this with previous experience. The aim was to allow it to fire not at the target itself, but at the place where the target should be by the time the shell arrives.

Weiner saw that the statistical study of time series data needed to be related to the basic task of communications engineering. This work led him to deal directly with the problem of how to couple a man to his machine. He realized that in humans, sensors tracked motion and allowed a person's body to correct for any errors in the way that muscles were operating via a series of feedback

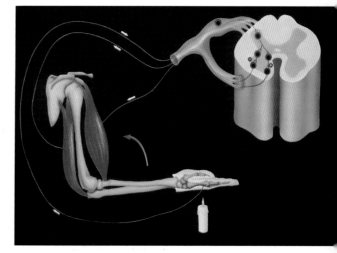

above *The human nervous reflex arc. The pain impulse passes via a sensory nerve from the hand to junctions in the white matter of the spinal cord (top right) which connect with two motor nerves that supply the biceps (right) and triceps muscles in the upper arm. Thus the sensation of pain caused by the heat of the flame provokes an upward jerk of the hand.*

mechanisms. Feedback was going to be the key to allowing machines to control themselves with accuracy. Furthermore, communication was clearly the cement of the nervous system, of society, and of any complexly organized structure.

His work during the war led to the 1948 publication of his book *Cybernetics, or Control and Communication in the Animal and Machine*. In this he described a way of looking at how the world functioned, based on the way that information is transmitted and processed.

For Weiner, the world was focused on information rather than energy, on digital or numeric processes rather than machine or analogue. This thinking led him to predict the future development of computers.

The book's impact spread far and wide. It introduced the now commonly used terms of "input," "output," and "feedback." For Weiner, communications engineering, the behavior of servomechanisms, of computing machines, and of the nervous system were all linked.

Keeping young

Weiner believed that mathematics was best performed by young people, once saying that, "Mathematics is very largely a young man's game. It is the athleticism of the intellect, making demands which can be satisfied to the full only when there is youth and strength." His own youth and strength, however, seemed to persist for most of his life and he made a continual stream of contributions to mathematics.

It came to an abrupt end on March 18, 1964, when he died just two months after being awarded the National Medal of Science in a ceremony at the White House in Washington D.C.

Alan Turing

1912–1954

Acquaintances
— Ludwig Josef Johann Wittgenstein
  (1889–1951)
— John von Neumann (1903–1957)
— Alonso Church (1903–1995)
— William Gordon Welchman (1906–1985)

From an early age, Alan Turing showed a desire to find his own solutions to problems—his answering of a philosophical problem in mathematics was the first step that led to the world of computers. His concepts were often innovative: when addressing a problem posed by the radical mathematician David Hilbert, he devised a solution based on a hypothetical machine. The machine obeyed instructions set out in an algorithm. The fact that Turing solved Hilbert's problem is totally overshadowed by the range of applications to which successors of his Universal Machine have been applied. While he used them to crack codes, the world now uses computers for just about everything.

Challenging from the start

At school, Turing followed a familiar pattern for great thinkers—he found conventional schooling at best uninteresting, at worst incomprehensible. His handwriting was scruffy, he struggled in English, and in mathematics he was more interested in chasing his own ideas than in solving the simplistic problems set by the teachers. He loved chemistry, but to the consternation of his teachers, tended to carry out his own experiments. His head teacher once wrote, "If he is to stay at Public School, he must aim at becoming educated. If he is to be solely a Scientific Specialist, he is wasting his time at a Public School."

By the time he arrived at Cambridge University in 1933, Turing was well aware of the theories presented by people like Bertrand Russell, who thought that logic was a solid foundation for mathematical truth. He also knew that other mathematical philosophers were questioning how truth could be captured by any formalism. After all, in 1931 Kurt Gödel had shown that true statements about numbers could exist that could not be proved by applying sets of rules—mathematics was necessarily incomplete.

Then in 1935, Turing learnt about a question posed by Hilbert. It was the question of decidability—the *Entscheidungsproblem*: is it possible to find a definite method for deciding whether any given mathematical assertion was provable?

To come up with an answer, Turing realized that the first step would be to define exactly what was meant by the word "method." He started by analyzing how people performed methodical processes. From this he was able to show that people could work "mechanically." In a stroke of inspiration, he expressed the concept in terms of a theoretical machine that could perform certain precisely defined elementary operations by reading and writing symbols on paper tape. Such a machine, he said, was quite capable of doing everything that would count as a "definite method."

In terms of modern language, this definite method would be called an algorithm, and it was a short step for Turing to answer Hilbert's question in the negative: it is not possible to develop a method that can decide whether all mathematical assertions are provable.

In April 1936 Turing wrote up his concept and conclusions and tried to get it published. Unfortunately, American logician Alonso Church had come to a similar conclusion and was fractionally ahead of Turing. In the end, Turing's paper was accepted, but he was forced to refer to Church's work when *On Computable Numbers with an Application to the Entscheidungsproblem* was eventually published in August that year.

The universal Turing machine

There was a critical difference between Church's and Turing's work. While Church appealed to contemporary mathematics to make his point, Turing had introduced a theoretical machine. As such he had created a foundation for modern theories of computation.

Even though the world was decades away from seeing its first computer, Turing conceived the idea of a universal machine. He realized that an algorithm could be employed to solve a problem, and you could build a machine that obeyed a particular algorithm. The major difficulty with this was that each new question would demand building a new machine.

But imagine, suggested Turing, writing each algorithm as a set of instructions using a standard code. A machine could interpret the code mechanically and produce its answer. There was the possibility that one machine could perform all possible tasks.

left *The Enigma encryption machine used by the Germans in World War II to send messages in code. This electromechanical machine used three rotors and a complicated algorithm to encode messages which were typed in on its keyboard.*

above *The "Colossus" computer, Bletchley Park, 1943, the world's first electronic programmable computer.*

In doing this, he had created the intellectual framework required to write a computer program. All that was needed was a leap in technology. Nine years later developments in electronics allowed his ideas to rise from the paper and become embodied in what are now thought of as primitive computing devices.

Cracking codes

While working in Princeton University, Turing built a cipher machine. This used electromagnetic relays to multiply binary numbers. When he returned to England around the start of World War II it was almost inevitable that the government would employ him in their code-breaking headquarters at Bletchley Park.

Turing used his ideas of producing mechanical means of making logical deductions to break the codes that the Germans were creating. From late 1940 onwards, he and colleagues had developed their "Bombe," which made reading of Luftwaffe signals a routine process. The more complex Enigma codes used in German naval communications were, however, much more of a challenge. Turing started to crack the system at the end of 1939, but it took until mid-1941 before there was any routine success.

With access to German intelligence, the war started to turn in the Allies' favor. Then the German military added additional layers of complexity. Engineers started installing a telephone system at Bletchley Park in the hope of increasing the speed of handling messages, but the engineers soon found themselves producing logic machines to help crack codes. The process worked, and the Germans again lost their advantage.

To bigger machines

By the end of the war, Turing had overseen the first digital electronic machines in the form of the Colossus machines. He spent many hours learning electronics and was keen to embody his theoretical Universal Turing Machine in electronic form. He didn't know it, but he wanted to invent a digital computer.

The National Physical Laboratory on the outskirts of London planned a computing project that they hoped would rival a similar U.S. venture, and Turing was invited to join the team. In early 1946 they started work on ACE, the Automatic Computing Engine. By October 1947, the project was getting nowhere and Turing returned to Cambridge. A few months later Turing was given another opportunity when a second computing project started up in Manchester, and June 1948 saw the first practical demonstration of Turing's computer principle. The world had changed.

Claude
Shannon

1916–2001

Acquaintances
— Mikhail Botvinnik (1911–1995)

Computers may provide humankind with a powerful tool to handle and process information. They would, however, have been useless unless someone stopped to work out what information is. Claude Shannon's theory of information works by separating the predictable from the unpredictable, saying that only unpredictable data is information. His next insight was to code that information in short sequences of the most basic numbers available — 1s and 0s. This allowed information to be handled by arrays of mechanical switches and also enabled it to be transmitted as a series of pulses along telegraph lines. In so doing he had moved from information to communication.

Communication starts with information

"The fundamental problem of communication is that of reproducing at one point either exactly or approximately a message selected at another point." This opening sentence in a landmark paper by Claude Shannon not only explains the issue involved in communication, but also acted as a guide for his working life. The paper, entitled *A Mathematical Theory of Communication*, not only introduced the concept of communication theory, but also analyzed the concept of information.

For Shannon, "information" is the part of a message, data set, picture, or group of sounds that is not predictable. Rolling a dice that had a 6 on each of its faces would not reveal any information, since the outcome is totally predictable. Such a view of information immediately shows a way of reducing the volume of data that you need to be stored or transmitted. If this dice is rolled 100 times, you don't need to record 100 6s, just say 100 x 6. This, after all, was the information in the sequence, and the approach "compresses" the data.

Similarly, sorting or transmitting a number like π does not require you to send the infinite set of numbers down the line — they are predictable if you send the appropriate algorithm that can calculate it at the other end. The numbers in π are predictable once you have the algorithm.

Shannon realized that if you flip a coin there is one of two uncertainties — heads or tails. The information from a sequence of flips could be coded as a series of values 1 or 0. For a more complex data set you may need four values, which would require two digits: 00, 10, 01, and 11. A code group of three bits in length permits eight combinations and four bits allows for 32 — sufficient to transmit a standard set of characters used to write English.

Once he started using this system it again became obvious how information could be compressed. Frequently-used letter combinations or words could be given codes of their own, saving the need of transmitting each character.

Moving the message

Communication theory is concerned with the best way to transmit this information. It also looks at what might go wrong with the signal during transmission that could cause it to be misunderstood by the receiver. Shannon was excited that any sort of message could be transmitted as a series of 0s and 1s, regardless of whether it was words, numbers, pictures, or sound, and demonstrated that all sources of information — telegraph keys, people speaking, television cameras — have a "source rate" that can be measured in bits per second. Communication channels, he said, have a "capacity" that again is measured in bits per second. Obviously information can only flow if the source rate does not exceed the capacity.

During his first-degree courses he studied Boolean algebra and symbolic logic. He realized that this was a good way of studying two-value systems.

In addition, his Master's thesis had shown that binary digits could be represented as switches. Digit 1 was the switch turned on — 0 was it turned off. Using these switches along with Boolean algebra, he showed that the information could be processed automatically by electrical circuits. This has become the backbone concept in electronic circuits and computers, as well as for communication systems. He coined the term "bit" for a unit of information — a "1" or a "0". All communication lines are now measured in bits per second.

Shannon also became aware that his system might simply be following nature, in the way that biological systems store and transmit information. He spent the summer of 1939 working with Barbara Burks at Cold Spring Harbor, and produced a doctoral thesis entitled "An Algebra for Theoretical Genetics." The four-letter code used in DNA. operates in many ways like a four-value version of Shannon's binary system.

Maybe invention was in his blood as, later in life, he discovered that his childhood hero, Thomas Edison, was a distant cousin. After all, Shannon had started designing communication systems while at school. He made a telegraph system linking his house to that of a friend's one mile away, making use of the barbed wire that ran around nearby adjoining fields.

A joker and inventor

Building machines became one of Shannon's favorite activities. It was part work, part amusement. In 1950 he built Theseus, a maze-solving magnetic mouse with copper whiskers. Controlled by a relay circuit, Theseus moved around a maze of 25 squares. Each time the maze was altered, the mouse would set off in search of some arbitrary goal. If the territory was left unchanged, Theseus could be dropped into it at any point. It would search around until it reached a position that it recognized and would then set off to its goal.

Theseus' brain was a circuit of 100 relays and its muscles were a pair of motors driving an electromagnet under the board. The magnet moved the mouse. Theseus was a visual

left *Theseus was Shannon's magnetic "mouse." Using what would now be considered to be a minuscule amount of electronics, Shannon enabled this mechanical rodent to find its own way around a maze.*

demonstration of the power of Boolean algebra operating via relay-driven switches.

However, a more trivial creation was his ultimate machine. This was a small wooden box with a single switch on the outside. Flick the switch and an ominous buzzing starts inside. The lid lifts and a hand reaches out, turns off the switch and returns to the box. The lid snaps shut and the noise ceases.

A private checkmate

In 1950, Shannon wrote a paper called "Programming a Computer to Play Chess." At the time it was a considerable challenge. While many programs now achieve this, most follow the strategy that he proposed. Then, when giving a lecture in Russia, he met World Chess Champion Mikhail Botvinnik, who also doubled as an electrical engineer and was interested in the idea of programming a computer to play chess.

The two talked at length, but the conversation was not helped by the poor means of communication at hand—a translator who did not understand chess or computers. In the end they sat down to a game of chess. Shannon took an early lead, but was defeated after 42 moves.

The eccentric Shannon gained a reputation for keeping to himself by day and riding his unicycle down the corridors at night. The unicycle had been a Christmas gift from his wife.

A colleague at Bell Telephone's laboratories, New Jersey, D. Slepian, once summed him up by saying, "Many of us brought our lunches to work and played mathematical blackboard games but Claude rarely came. He worked with his door closed, mostly. But if you went in, he would be very patient and help you along. He could grasp a problem in zero time. He really was a genius. He's the only person I know whom I'd apply that word to."

Timeline

1916
Born in Petoskey, Michigan, on April 30

1936
Receives two Bachelor's degrees at the University of Michigan, a B.Sc. in electrical engineering and a B.Sc. in mathematics, and moves to the Massachusetts Institute of Technology (MIT)

1940
Receives his Master's degree with a thesis entitled "A Symbolic Analysis of Relay and Switching Circuits," and works as a National Research Fellow at Princeton University

1941
Joins the staff at the Bell Telephone laboratories in New Jersey

1948
Publishes *The Mathematical Theory of Communication*

1949
Marries Betty Moore. The couple subsequently have four children— Robert, James, Andrew, and Margarita

1956
While still officially working for Bell Telephone, Shannon becomes the professor of communications science and mathematics at MIT

1958
Becomes the Donner Professor of science at MIT

2001
Dies in Medford, Massachusetts, on February 24

A process of slow change that has been occurring over a long time is suddenly interrupted by a rapid violent event—a catastrophe. René Thom developed a mathematical approach that allowed people to analyze such systems and offered the possibility of making predictions about the likelihood of a catastrophe or how to take steps either to cause it or to prevent it occurring. Although Thom became disenchanted by the theory toward the end of his working life, it presented a new way of analyzing natural events in fields as varied as engineering and the social sciences.

René Thom

1923–

Acquaintances
— Albert Einstein (1879–1955)
— Herman Klaus Hugo Weyl (1885–1955)
— Henri Cartan (1904–)
— Charles Ehresmann (1905–1979)
— Norman Earl Steenrod (1910–1971)
— Kunihiko Kodaira (1915–1997)
— Alexander Grothendieck (1928–)

Moths with appetite

Many plantations of spruce trees grow in northern America. The trees are large and the wood is valuable, particularly in the case of those over 60 years old. They have a predator. These mighty trees can be ruined by a delicate moth and its hungry larvae. The spruce budworm (*Choristoneura fumiferana*) strips the leaves off the spruce and particularly thrives on older trees.

Its preference for older trees causes an interesting phenomenon. As a forest grows, the number of moths stays fairly static, and causes little trouble. Then as some trees get to over 74 years old, any weather event that enables successful breeding can lead to an eruption in the population of moths. Within weeks the forest is stripped of its leaves.

When René Thom started looking at the mathematics of the way that continuous actions can result in discontinuous results, it is fair to say that he wasn't thinking about *C. fumiferana*. But the concepts that he identified have allowed people to model these natural situations and generate preventative strategies. In the case of the forest, the best thing to do is to reduce the average age of tree to keep below critical levels.

Catastrophe—a theory of abrupt change

Since his early days as a student, Thom had been interested in problems of topology—the study of three-dimensional shapes. His undergraduate life was spent in Nazi-occupied Paris, but intellectually it was an exciting time for Thom as he began to work alongside French mathematician Henri Cartan.

Cartan encouraged him to look at the work that American mathematician Norman Earl Steenrod was doing on the topology of spheres and cubes, as well as Steenrod's theories about the mathematics of fiber bundles. As World War II came to an end, Thom and Cartan both moved to Strasbourg, but Thom chose to present his Ph.D. thesis in Paris.

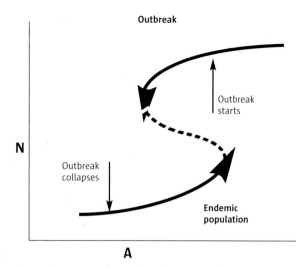

above **Catastrophe theory shows how a population can either explode in an outbreak or fall rapidly as the outbreak collapses.**

With his Ph.D. complete, Thom was awarded a fellowship that enabled him to travel to the United States in 1951, and allowed him to meet Albert Einstein, Herman Klaus Hugo Weyl, and Norman Earl Steenrod, and attend seminars given by Kunihiko Kodaira. He returned and taught for a short time in Grenoble, France, before returning to Strasbourg. He was awarded a professorship, but then he moved to the Institut des Hautes Etudes Scientifiques, just south of Paris. This turned out to be a stressful episode for Thom, because he felt that he was constantly living under the shadow of Alexander Grothendieck and that in comparison to Grothendieck he had nothing new to offer.

The pressure caused Thom to redirect his efforts away from strict mathematics, to what he once described as "a very general form of 'philosophical' biology." The major method in this theory is to distinguish between aspects of a system that change slowly over time, and those that can change rapidly. In the spruce and moth example, the trees change slowly, but the moth population can suddenly explode.

This was the period of his work when he generated catastrophe theory, where catastrophe refers to a loss of stability in a gently changing, or dynamic, system. Thom characterized many different forms of catastrophe, depending on the number of different elements involved. The simplest has two components and produces the classic "cusp" catastrophe. In his 1972 book *Structural Stability and Morphogenesis*, Thom applied his concepts to analyze the growth of embryos, and it has also been used in modeling nerve transmission and heartbeat.

Various people have applied the theory to phenomena such as the stability of ships, load-bearing capacity of bridges, wars, and riots in prisons. In his autobiography, Thom says, "As soon as it became clear that the theory did not permit quantitative prediction, all good minds ... decided it was of no value." All the same, Thom's catastrophe theory gave a fresh insight into the way that mathematics can influence all other branches of science.

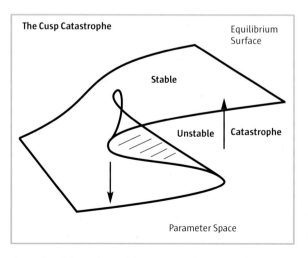

above **René Thom showed that catastrophes occur when you rise or fall from one stable state to another.**

By the mid-1900s mathematicians were used to dealing with four dimensions—lines, planes, solids, and time. While physicists revealed further dimensions in their quest to understand quantum dynamics and the working of atoms, Benoit Mandelbrot started to probe the regions between the first four. He claimed that the real world in which we live was best described by his fractal geometry, a geometry that finds space between the main dimensions. The simple algorithms used in this science have revealed an understanding of objects from snowflakes to mountain profiles, from trends in stock market prices to pretty pictures with infinite detail in their pattern.

1924–

Acquaintances
— Paul Lévy (1886–1971)
— Norbert Wiener (1894–1964)
— John von Neumann (1903–1957)

Benoit Mandelbrot

A fresh look finds fractals

Prior to Benoit Mandelbrot, geometry was based on the Euclidean understanding of space. Living in Greece at c.300 B.C., mathematician Euclid had written 13 books, some of which dealt specifically with geometry. They had formed the foundation of the concept of three physical dimensions, and little had occurred over two millennia to fundamentally rock that foundation.

Then came Albert Einstein, who introduced a fourth dimension—time. Next up was Benoit Mandelbrot, who made his great discoveries by defying established academic mathematics. He went beyond Einstein's theories to discover that the fourth dimension includes not only the first three dimensions, but also the gaps or intervals between them, the fractal dimensions. In so doing he showed that mathematics could have a valuable role in describing the workings of nature.

Mathematics was in Mandelbrot's blood and upbringing. His uncle, Szolem Mandelbrot, was a member of an elite group of French mathematicians in Paris known as the "Bourbaki." Benoit Mandelbrot was born in Warsaw in 1924 to a Lithuanian Jewish family, but his family moved to Paris in 1936 to seek protection from persecution. They may have been safer, but young Benoit received no formal education.

After the war Mandelbrot flourished at mathematics, but preferred to visualize questions, rather than work through established systems of logic, and after getting a Ph.D. he set off to the United States in the hope of finding some intellectual freedom.

Eventually, in 1958, he started work at I.B.M.'s research center in Yorktown Heights, New York, where he pursued his mathematics at the same time as playing with the world's biggest computers. For a few years he defied the academic concept of specializing in a particular area, and instead dabbled in linguistics, game theories, aeronautics, engineering, economics, physiology, geography, astronomy, and of course physics.

Every now and then, his work did spark some distinct interest. He started looking at the day-to-day changes in the price of cotton, a commodity for which records went back a couple of hundred years. Eyeballing the data showed nothing of interest, but analyzing it with computers revealed overall patterns in the price movements. This was shocking. If there were patterns, then stock-traders would be able make accurate predictions of what could occur in the future. At this point, though Mandelbrot didn't realize it, he had stumbled upon fractals.

The Mandelbrot set

In 1975, Mandelbrot published *Les objets fractals*, a book that set out his new idea of fractal geometry. For all its complexity, the whole concept revolves around a remarkably simple formula: $z_1 = z_0^2 + c$. The idea is that you repeat the calculation time and again, starting with a specific value of z_0 and then replacing z_0 with z_1 in each subsequent calculation. The value c is the same as the original z_0, and remains unchanged at each repeat.

Mandelbrot showed that making c a real number, such

above **A detail from the Mandelbrot Set—a form of fractal geometry— plotted from complex number coordinates.**

as 1, 2, 3, or 4, produced patterns, but the fun started with complex numbers, like 1.1, -1.1, and -1.38. If the process was repeated, or iterated, millions of times, the resulting number is always between 2 and −2. Looking at the developing sequence shows that patterns start to occur. These mathematical patterns can be represented visually on computer screens as ever-unfolding snowflake-like designs, which give the appearance of continuously revealing more and more detail. The group of numbers creating the picture is a Mandelbrot set.

Mandelbrot's formula allowed him to start to characterize features like the shape of a cloud, a mountain, a coastline, or a tree. As Mandelbrot said in his book *The Fractal Geometry of Nature* (1983): "Clouds are not spheres, mountains are not cones, coastlines are not circles, and bark is not smooth, nor does lightning travel in a straight line." Before this work mathematicians believed that most of the patterns of nature were far too complex, irregular, fragmented and amorphous to be described mathematically. But Mandelbrot's new fractal geometry of nature, based on the fourth dimension and complex numbers, was capable of describing the most amorphous and chaotic forms of the real world.

Nature between dimensions

If the four dimensions of geometry had seemed like a safe climbing frame for entertaining intellectual exercises, Mandelbrot was about to show that the real world was far more complex. He invented the term "fractal dimension" from the Latin adjective *fractus*, a word derived from *frangere*, meaning to break or to create irregular fragments.

Fractals help us to see nature in a new light. For instance, look at the irregular shape of a mountain and then look closer at a small part of the mountain—the same basic shape of the whole mountain repeats itself again on a smaller scale. Look closer still and the same shape is there. And so on to infinity. This happens within the pictures produced by Mandelbrot sets, where an infinite number of smaller Mandelbrot shapes hide within the zigzagged, spiraling edges of the overall form.

Living in the real world involves competition and cooperation. In situations like warfare, the competition is pure and the cooperation between opponents is non-existent. In most activities, however, people cooperate and find the solution that presents the best outcome for all participants. John Nash applied games theories to situations as varied as wars, stock markets, and evolution. The key to this was developing the concept of an equilibrium point — a solution that maximizes everyone's benefits. He has had to fight mental illness, and has come to his own equilibrium point where he is mentally stable, but at the same time not as creative as he would wish to be.

John Nash

1928–

Acquaintances
— Solomon Lefschetz (1884–1972)
— Norbert Wiener (1894–1964)
— John Synge (1897–1995)
— Emil Artin (1898–1962)
— Oskar Morgenstern (1902–1976)
— John von Neumann (1903–1957)

above Nash's influence spread to many areas involving the interaction between people, but it has probably had most impact in economics where his ideas have affected the way people trade and do business.

Games and economics

While working for his first degree at Carnegie Institute of Technology, Pittsburgh, John Nash took an elective course in international economics. It was an unusual choice for someone otherwise majoring in chemistry and mathematics. The course, however, stimulated his imagination about the rules involved in bargaining and led him to write *The Bargaining Problem*. This marked him as an unusual student, as not many publish an academic paper while still studying for their undergraduate qualifications.

The subject intrigued Nash and led him to look at John von Neumann and Oskar Morgenstern's work on game theory. This is an area of mathematics that looks at how people behave when placed in competitive situations.

Up to this point the research was restricted to looking at situations where there are two combatants and the result is that one wins and the other loses. In math jargon, this is a zero-sum game. While the work had its uses in warfare, it had only limited value to helping people operate in more normal situations.

Nash's grounding in economics had taught him that people normally operate so that they reach a position of mutual benefit, and he developed systems for understanding these non-zero-sum situations. He also pointed to the differences that exist between situations where competitors sit together and cooperate to produce an outcome, and those where there is no cooperation. To sort this out he introduced the concept of an equilibrium point—a collection of the various players' strategies where no individual player can improve his or her outcome by changing strategy.

Schizophrenia and the Nobel Prize

His ability to prove that there will always be such a point of equilibrium has had such an impact in economics that he was nominated for the Nobel Prize for Economics. The Nobel committee, however, had a dilemma. They only give prizes to people who have done outstanding work, and who are likely to continue doing outstanding work in the future. But

by the time that Nash's work was recognized he was suffering from paranoid schizophrenia.

Norbert Wiener was one of the first people to detect that Nash's eccentric patterns of behavior were the symptoms of mental illness, and in the spring of 1959—with his wife pregnant, and Nash himself apparently on the threshold of a glittering career—Nash resigned his job. After spending 50 days at the McLean Hospital, he set off to Europe, attempting to gain entry under the status of a refugee. The next two decades saw him spend time in and out of hospitals, always, he claimed, held there against his will.

Then, in the early 1990s Nash found ways of avoiding the symptoms of schizophrenia and resumed his work. It was at this point that the Nobel committee placed him on their list of potential recipients, and in 1994 he was duly awarded the Prize for Economics.

From Nash's point of view a return to stable health had mixed blessings. In the autobiography that he wrote to accompany the Nobel Prize, he contrasts his state with that of someone who had recovered from a physical disability and was now in good health. He, however, was mentally stable, but that very stability limited his ability for creative thinking. And Nash has plenty to think about, because, while his work on game theory has had the most impact on global affairs, his work on other subjects like geometry and topology has arguably had the greater effect on mathematics.

Timeline

1928
Born in Bluefield, West Virginia, on June 13
1948
Receives a B.A. and M.A. from the then Carnegie Institute of Technology, Pittsburgh, and moves on to Princeton University
1949
While studying for his Ph.D., Nash writes a paper on non-cooperative game theory that 45 years later will win him a Nobel Prize for Economics
1950
Receives his Ph.D. His thesis is entitled "Non-cooperative Games"
1952
Nash publishes his paper *Real Algebraic Manifolds* and moves to teach at the Massachusetts Institute of Technology. Here he meets Eleanor Stier and in 1953 they have a son, David Stier
1954
Nash is arrested in a police operation to trap homosexual men
1957
Nash marries Alicia Larde, originally from El Salvador
1958
Alicia becomes pregnant, and Nash develops paranoid schizophrenia
1994
Nash is eventually awarded the Nobel Prize for Economics
1996
After many years of mental illness, Nash recovers enough to present a paper at the tenth World Congress of Psychiatry

Good science demands that people observe the world around them with care and make accurate records of what they discover. Few, if any, have done this better than Leonardo da Vinci. For him, seeing was the main avenue to knowledge, and as such he trained his skills as an artist. While he is widely known as a painter, the bulk of his surviving work is manuscripts of mechanical, architectural, and anatomical drawings, in which he was keen to demonstrate how each object performed its task. He combined observation with invention and designed many fanciful machines, including a prototype airplane and helicopter.

Leonardo da Vinci

1452–1519
Acquaintances
— Antonio Pollaiuolo (1429–1498)
— Andrea del Verrochio (c.1435–c.1488)
— Lucas Pacioli (c.1445–c.1514)
— Ludovico Sforza (1451–1508)
— Marcantonio della Torre (1473–1506)
— Giovanni Francesco Rustici (1474–1554)
— Cesare Borgia (1476–1507)
— Francesco Melzi (c.1490–1568)

Seeing and thinking

Renaissance Italy was a turbulent place, with frequent wars and changes of ruler. It was also a place where rulers were interested in gaining technical advantage over their foes, and keen to excel in art as well as in war. It was therefore a superb environment for Leonardo da Vinci's enquiring mind. The illegitimate son of Ser Piero di Antonio, a landlord and notary, and peasant woman Catarina, Leonardo spent his childhood living in his father's household. Ser Piero appears to have welcomed the child into his home after his marriage to Albierra di Giovanni Amadori failed to result in any children.

As an apprentice in Florence, Leonardo had the task of mixing pigments for paints as well as assisting artists cast metal statues. He also studied anatomy as he practiced drawing artists' models. It was an ideal tutorship in developing the technical skills he would employ throughout his life.

It would appear that he saw no subdivisions between art, science, and engineering. While other intellectual artists were developing mathematical principles for painting and laws of perspective, Leonardo took the process further. His aim was to use his eyes to see how things are made, and to record them in ways that make it clear how they work.

He wanted to observe all objects in the visible world and determine their structure. For him, art had become a tool for discovery and invention. This applied as much to his detailed anatomical drawings of the heart as it did to technical illustrations of war machines designed to hurl rocks at enemy fortifications.

Body and machine

Leonardo's science became the key feature of his work while he was in Milan. There he started making copious notes of his thoughts and observations. At first he appears most concerned about developing a theory of art, and as part of this he contributed the now famous drawings of a symmetrical human body contained within a circle and a square in Lucas Pacioli's book *On Divine Proportion*, that was published in 1494.

Moving to Florence, Leonardo started dissecting bodies in the hospital of St. Maria Nuova, concentrating on not only recording the shape and position of organs, but also determining their function from an analysis of their structure. During his life he worked on 30 bodies. At the same time he became fascinated with the flight of birds, and some of his drawings show how he envisaged building a human-powered flying machine.

In drawing machines, he was keen to show how individual components worked. Like a child's construction set, this allowed him to recombine the components to create novel devices.

He saw air as a material that had many properties similar to water, and was interested in the ways that currents moved through both media. His diagrams of pumps and Artesian screws show how he developed an understanding of the ways that machines could interact with air and water, and his diagram of a prototype helicopter works on the basis of it being a screw that would drive itself upward into the air.

His inventiveness led to further discoveries. For example, in a note written in 1513, now called Codex Arundul, Leonardo says, "… in order to observe the nature of the planet, open the roof and bring the image of a single planet onto the base of a concave mirror. The image of the planet reflected by the base will show the surface of the planet much magnified." From his moon gazing he concluded that it shone because it reflected sunlight, but he made the mistake of believing that the Moon, like the Earth, had seas and areas of dry land.

Leonardo da Vinci's legacy goes beyond his drawings. Leonardo showed how a spirit of inquiry, coupled with diligent observation, can reveal the way that natural systems operate, and he employed these principles in his creation of buildings and machines.

Timeline

1452
Leonardo da Vinci is born at Anchian near Vinci, Tuscany, on April 15

1476
An anonymous allegation claims that Leonardo is having a homosexual relationship with artists' model, 17-year-old Jacopo Saltarelli. The charge is dismissed

1482
Enters the service of Ludovico Sforza, Duke of Milan, as *pictor et ingeniarius ducalis* (painter and engineer of the Duke)

1484-1489
Works on architecture, military and hydraulic engineering, flying machines, and anatomy

1492
In the year that Columbus reaches the New World, Leonardo visits Rome

1499
Three months after the French invade Italy, Leonardo travels to Venice and Florence

1502
Pope Alexander VI's son, Cesare Borgia, employs him as "Senior military architect and general engineer" [Chief engineer]

1503
Begins painting the Mona Lisa

1505
Studies bird flight and geometry

1508
Studies anatomy in Milan

1509
Draws maps and geological surveys of Lombardy and Lake Isea

1516
Moves to France, to work for Francois I, although his right hand has been partially paralyzed by a stroke

1519
Dies on May 2 at the Castle of Cloux near Amboise, France, and is buried in the Church of St. Florentine, but his remains are scattered during the Wars of Religion

Steam and the industrial revolution came hand in hand. The invention of industrial manufacturing processes created a need for a controllable and confinable source of regular power—the invention of steam engines enabled people to design new machinery. James Watt was in the right place at the right time. As a capable technician and inquiring inventor he saw limitations in the early machines that employed steam and he came up with solutions. He therefore can be seen to stand at the junction of basic science, which had discovered the existence of atmospheric pressure, and technological progress, which started to capture the power of that pressure.

James Watt

1736–1819

Acquaintances
— John Anderson (1726–1796)
— Matthew Boulton (1728–1809)
— William Murdock (1754–1839)
— William Symington (1763–1831)

Right place, right time

Born in Greenock, Scotland, on January 19, 1736, James Watt was the son of a successful ship's chandler. Having poor health as a child, Watt had little formal education. He did, however, enjoy pottering around in his father's workshops building tools and repairing anything that came to hand. One, possibly proverbial, story tells of Watt's grandmother becoming annoyed that he could spend an entire afternoon messing about with the steam that came from a kettle.

In 1757, Watt got a job at the University of Glasgow, where he was employed to make scientific instruments. One of the university's instruments was a model of a steam-powered pump, originally designed by Thomas Newcomen. Newcomen was interested in pumping water out of mines. He built a pump that had two cylinders with pistons inside. One had a pipe running down into the mine, and the other could be heated and filled with steam. A large pivoting beam connected the two cylinders. When the steam-filled cylinder was cooled, the steam condensed and the piston was drawn down, operating the pump.

A key limitation to the pump's efficiency was that each time you wanted to run it you had to use a lot of energy to heat up the vessel. This meant that it was slow and wasteful. But one day, Scottish scientist John Anderson asked Watt to repair the university's Newcomen pump, and as he worked Watt thought of a way to improve it.

Watt developed an engine that used two cylinders connected by a pivoting beam. In one cylinder the piston was driven by steam, and the connecting beam pulled the other piston, which pumped water from the mine. The stroke of genius, however, was in the way that Watt got the maximum efficiency. He had found an ingenious way of using the steam twice.

To start with, the steam pushed the piston down to the bottom of the cylinder. At this point a valve opened allowing the steam to pass from above the piston to below. This allowed the piston to return to its starting position. Watt realized that if he could condense the steam beneath the piston, he would create a vacuum which would pull the piston down. So at the same time as letting new steam in above the piston, he opened a valve in the lower part of the cylinder connecting this steam to an underground condenser. The piston was therefore pushed down by steam and pulled down by the vacuum.

A machine for all uses

The increase in efficiency made Watt's steam engine attractive for all sorts of uses. It could generate power to motivate machines in factories. This was the late 18th century—the start of the industrial revolution. Machines were moving into the manufacturing industry. Up to this point, any factories that needed a source of power were largely dependent on wind power or waterwheels.

It is an interesting debate as to whether steam generated the industrial revolution, or the revolution generated steam power. In *Das Kapital*, Karl Marx gives his opinion, describing how the industrial revolution brought about the invention of steam power rather than the other way around.

Which ever came first, steam engines gave the ability to generate power inside cities. Manufacturing industry was no longer dependent on wind or water for its power and no longer needed to be spread between isolated villages, each of which ran its own little mill. Aware of the universal application of his machine, Watt was careful in his 1784 patent application to talk of his invention not as something designed for a specific purpose, but as a machine that had universal use in industry.

The legacy

New machines were soon appearing and each one was lighter and more powerful. This increasing power to weight ratio opened new opportunities, and in 1788 William Symington took a steam-powered catamaran across Dalswinton loch. By 1819, the year that Watt died, there were 18 steam-powered weaving factories in Glasgow, containing a staggering 2,000 looms.

Sixty-three years after Watt's death, the British Association gave his name to the unit of power—the watt.

Timeline of steam

1600s
Evangelista Torricelli (1608–1647) and Vincenzo Viviani (1622–1703) realize that the weight of air pushing on a reservoir of mercury can force the liquid to rise into a tube that contains no air, that is, that contains a vacuum

1650
Otto von Guericke (1602–1686) invents an air pump, and shows that if you removed the air from the center of two hemispheres that were resting together, the pressure of the outside air meant that 16 horses were not powerful enough to pull the hemispheres apart

1698
Thomas Savery (c.1650–1715) makes a steam pump that can pull water up from mines. He built a large vessel and ran a pipe from it down into a mine. He filled the vessel with steam and sealed it. Then, as the steam condensed in the vessel it caused a partial vacuum. Water was drawn up from the mine into the vessel, the pressure of the atmosphere pushing down on the water in the bottom of the mine and forcing it up the tube into the evacuated vessel

1712
Thomas Newcomen (1663–1729) uses the property of condensing steam to create a partial vacuum in a cylinder and therefore pull a piston. The system was highly inefficient, but was used to pump water out of mines

1769
Watt obtains a patent for improving Newcomen's design

1784
Watt patents his double-acting steam engine

1804
Richard Trevithick (1771–1833) builds the world's first steam-powered locomotive

1819
Dies August 25

Some geniuses have caused us to take a fresh approach in our philosophical view of life. Thomas Alva Edison had a different effect — he has radically altered the way that we see our physical world. His invention of the electric light bulb, combined with the setting up of companies to supply homes with electricity, has led to a 24-hour economy and work patterns that are not dependent on nature's supply of light. By inventing cameras that could record moving images, he enabled people to sit in cinemas and in effect travel the world — to witness current affairs with a heightened sense of reality.

Thomas Edison

1847–1931

Acquaintances
— Eadweard Muybridge (1830–1904)
— George Westinghouse (1846–1914)
— Edwin Stratton Porter (1870–1941)

From rails to lines

There is often a fine line between a scientist and an inventor. The scientist probes the world to see how it works, and the inventor makes use of the findings in creating machines, instruments, and tools. Thomas Edison started life as a scientist, who found that he kept inventing. By the end of his life he had filed a staggering 1,093 patents.

He wasn't marked out for greatness at birth. His father worked in the timber trade at Port Huron, Michigan, and Edison went to the local school. Or rather, he sometimes went to the local school. He had hearing difficulties, rapidly became bored, and often played truant. His mother believed in his ability, and allowed him to set up a smelly laboratory in their basement in which to carry out his experiments.

In 1859, at the age of 12, Edison took a job selling newspapers and sweets on the Grand Trunk Railway and moved his laboratory to a disused railway carriage. When Edison saved a station official's three-year-old child from an oncoming train, his life changed. As a reward the official, J.U. MacKenzie, taught Edison how to operate the telegraph, and he never looked back.

From 1862 to 1868, Edison traveled around the United States and Canada working as an operator, but also developing new instruments that could improve the system. Convinced that he could make a life as an inventor, he set up a small business and filed a patent for an electric vote counter. It worked well, but no politician was interested in it. Edison had learnt a vital lesson. Don't invent things that people don't want to buy.

Moving to New York, he had the good fortune to mend a stock-counting machine at Samuel Law's Gold Indicator Company. Management were so impressed they employed him to sort out all of their machines. His reward was enough money to set up a serious designing laboratory.

Inventing communication

In the 1860s and 1870s electricity was new on the scene, and regarded by most people as a mere novelty. Edison saw it as a utility, a product with massive potential. He had seen the way it worked for transmitting telegraph messages, and developed carbon disks that would act as microphones and speakers in telephones. People who were miles apart could hear each other—the world had begun to shrink.

He then explored ways of recording sound, first on tin foil and then on wax-coated drums. He also developed ways of recording moving images on reels of film, though his attempt at combining the two, in the form of his Kinetophone, was less successful.

It was after seeing an exhibition of a 500-candlepower arc light that Edison boldly announced he would invent a safe and inexpensive electric light that would replace the gas lights currently lighting millions of homes. On October 21, 1879, Edison unveiled his carbon-filament lamp, which was supplied by power from his special high-voltage dynamos.

Working on a means of recording moving images had given Edison another way of shedding new light on the world, and as soon as his system was good enough he established studios to bring fiction to the masses.

The Edison effect

Edison continued to experiment with his light bulbs. In 1883 one of his technicians found that, in a vacuum, electrons flow from a heated element—such as an incandescent lamp filament—to a cooler metal plate. Edison saw no special value in the effect, but he patented it anyway. It subsequently became known as the Edison effect.

The Edison effect has an interesting feature. The electrons can flow only from the hot element to the cool plate, but never the other way. So when English physicist John Ambrose Fleming heard of the Edison effect he realized that an Edison-effect lamp would solve his problem of turning an alternating current into a direct current; he called the device a valve. In modern electronics the valve has been replaced by diodes, but the principle is still used in standard television screens and X-ray machines. Once again, Edison's work was to be involved in giving us new views of our world and ourselves.

Timeline

1847
Thomas Alva Edison is born on February 11 in Milan, Ohio

1871
Marries Mary Stilwell on December 25. They have one daughter and two sons

1877
Invents carbon telephone transmitter and files a patent for a tin foil cylinder phonograph

1879
Devises an electric incandescent light bulb

1880
Discovers the "Edison effect"

1882
Opens a commercial electric station in New York City with 85 customers

1884
Edison's wife, Mary, dies on August 9

1886
Marries Mina Miller on February 24. They have one daughter and two sons

1887
Develops a wax cylinder phonograph

1893
Constructs a film studio

1894–95
Develops the Kinetophone, which loosely synchronizes images with sound

1898
Spanish-American War occurs; Edison Company sends cameraman to Cuba to film scenes of war

1901
Develops process for mass-producing duplicate wax cylinders

1928
Edison is awarded Congressional Gold Medal

1931
Edison dies in West Orange, New Jersey, on October 18

Up until the end of the 19th century there were essentially two main types of people — a few who had money and a great many who hadn't. Henry Ford altered this balance and created large middle-class populations with money and time to spend. He did this, not so much by employing conveyor-belt-driven mass production, but by engineering the economic environment to create mass consumption. At the same time as establishing factories that could produce high-quality inexpensive cars, he paid his workers enough money so that they could buy one. In one stroke he had provided supply and demand.

Henry Ford

1863–1947

Acquaintances
- John Burroughs (1837–1921)
- Thomas Alva Edison (1847–1931)
- Luther Burbanks (1849–1926)
- Warren Gamaliel Harding (1865–1923)
- Harvey Samuel Firestone (1868–1938)
- Herbert Clark Hoover (1874–1964)
- Charles Augustus Lindbergh (1902–1974)

Have money, will spend

The man who produced an affordable car and created an urbanized middle class was born on a farm near Dearborn, Michigan, on July 30, 1863, and educated in the local schools. He loved building machines and in 1903 founded the Ford Motor Company, but it took another decade before Henry Ford changed the world.

In 1908 he unveiled the Model T, but it was not until 1913 that Ford began using standardized interchangeable parts and assembly-line techniques. They were not his ideas as mass production was becoming widespread — he simply adopted them rigorously. The immediate consequence was almost an end to the company. The laborers found the work unpleasant and monotonous and staff turnover soared to between 40 and 60 percent each year.

He worked out that if he paid his factory-workers a high wage, and produced more cars in less time for less money, then everyone would have enough money to buy a car. He famously said, "I will build a motor car for the great multitude ... constructed of the best materials, by the best men to be hired, after the simplest designs that modern engineering can devise ... so low in price that no man making a good salary will be unable to own one and enjoy with his family the blessing of hours of pleasure in God's great open spaces."

In 1914, Ford introduced a minimum wage of $5 a day. This was shocking. The average wage in the automobile industry was about $2.40 a day. Furthermore, Ford reduced the working period from nine hours to eight. The Wall Street Journal called it "economic crime," and as he then daily increased the wages towards $10, "Fordism" was scornfully used as a term of recklessness. But Ford became convinced. Pay them more and they will buy more — find ways of manufacturing more cheaply and the masses will flock to the shops.

People flocked to Detroit, hoping to work in one of Ford's factories, and by the end of 1914, the world's first automatic conveyor belt at the sprawling Highland Park plant was producing one car every 93 minutes. By 1926 the company had built some 15 million cars.

His vision would stimulate a middle class, and accelerate the pace at which people left the countryside and moved to towns and cities. When Ford left his family farm in 1879, aged 16, less than 20 percent of the American population lived in towns. By the outbreak of World War II this had risen to around 40 percent.

Ford was also acutely aware that the cars needed an infrastructure of support. History had established that every settlement had its blacksmiths, and Ford realized that in a similar way, every community needed fuel stations. There wasn't time for history to let these evolve — if the market was to be created, active effort was required across the country. Consequently, he invented the concept of dealer-franchise systems to sell and service cars.

In a stroke of paternalism, he set up a Sociological Department to try and teach people how to spend this new-found wealth sensibly. He was anxious that it wouldn't be wasted on vice and alcohol, and banned smoking from his factories. Apart from anything else, if people were spending too much money on these items they would have none left to buy cars.

A friendship with giants

Shortly after marrying Clara Bryant, Ford started working as the chief engineer at the Edison Illuminating Company, and through this he became a friend and business partner of the great inventor Thomas Alva Edison. After starting to manufacture cars, Ford had asked Edison to invent a means of storing electricity in a car, and though Edison worked on the project for some years, he never came up with a workable solution. It was Edison who convinced Ford to use a petrol-powered engine, rather than Ford's preferred option of an electric motor.

Their friendship extended beyond work, in that Ford and Edison regularly joined up with naturalists John Burroughs and Luther Burbank for camping expeditions. On occasion they were joined by Harvey Samuel Firestone and the President of the United States, Warren G. Harding.

On doctors' advice Edison moved from New Jersey to the warmer climes of Fort Myers, Florida. Ford was a frequent visitor and in 1916 he bought the house next door. When Edison needed to use a wheelchair, Ford bought one for himself so that the two companions could have races. Four years after the death of his only son Edsel, Ford died on April 7, 1947, in his bed, having created not just cars, but a mass market and a money-spending middle class.

opposite *Henry Ford (left) and Thomas Edison (right) were good friends, Ford having worked as the Chief Engineer for the Edison Lighting Company.*

left *Henry Ford sitting on the first car he ever built back in 1896 during his spare time. The quadricycle had two cylinders, was rated at four horsepower and had forward gears only.*

People have always recorded their thoughts on physical materials, such as cave walls, clay tablets, or paper. In each case, the materials used have influenced not only the message given, but also the thought-processes of the author or artist. Theodor Holm (Ted) Nelson saw that using paper constrained people to a linear train of thought. Before personal computers had arrived on the scene, he realized that this imminent technology could remove that constraint. By introducing the concept of hypertext he challenged people to take a more radical look at the way they develop ideas and to have a multi-dimensional approach to storing and processing information.

Ted Nelson

1937–

Acquaintances, contemporaries and influences
— Vannevar Bush (1890–1974)
— Marshall McLuhan (1911–1980)
— John Cunningham Lilly (1915–2001)
— Timothy Leary (1920–1996)
— William Jovanovich (1920–2001)
— Douglas Engelbart (1925–)
— Tim Berners-Lee (1955–)

From linear to multiple-parallel

Many people have a simplistic view of written communication, in which they see the author as the active person who creates the work and the reader as the passive recipient. They also assume that a reader always starts at the beginning and finishes at the end. In the 1960s, while studying sociology at Harvard University, Boston, Ted Nelson realized that this was not the case, and that computers were about to add new opportunities and complexities to the task of reading.

He had been brought up largely by his maternal grandparents, while both of his parents pursued careers in film, television, and on the stage. From this he developed a love of the stage, and when his father took him into television studios he realized that media are not static, but constantly changing. He was later to perceive that with the newly available technologies, film and literature could be presented not merely in a preordained sequence, but in multiple parallel versions.

Even before he arrived in the intellectual hothouse of Swarthmore College to study for a degree, he was dissatisfied with paper. Recording ideas on paper limited his ability to keep track of changes that he made to the text and restricted his power to create multiple links within the document and between different documents. Adding footnotes to a document was a beginning, but was not very powerful. He was also aware of the work of people like Vannevar Bush and Marshall McLuhan, who were challenging our ideas of communication and maintaining that the medium of communication is itself part of the message.

Enrolling on a computer course at Harvard he saw a potential solution. The computer course was the resolution between his interest in show business and academia — his desire to communicate vividly and his interest in the philosophy of communication.

As the second inventor of word processing, after Douglas Englebard, he saw that computers would soon allow people to revise drafts of their work with great ease. Nelson became convinced that most future creative work would take place on computer screens, and he felt that it was his job to create the systems and document structure that would underpin this activity. This both excited and worried him: word processing would enable a new approach to writing, but it would erase the record of the thought-processes that had generated the final document. He was also keen to see a document structure that encouraged intercomparison. For example, in the case of a map of the United States, the document would enable the reader to flick between ancient and modern maps, or geological and political maps, and thus gain a fuller understanding of the subject. In addition, he wanted documents to show the sources of their material, so the readers could follow the information trail for themselves.

Nelson realized that holding text on computers would enable readers to choose their own path through the document, or move to other documents whenever they wanted. The reader as well as the author would now play an active role in deciding where to go next.

He compared this process to an enormous version of footnotes or captions used in conventional publishing. In his concept of a Xanaolgical structure for a Docuverse, the world would contain one huge, interconnected electronic library, where both authors and readers could create links between documents, make changes, and add comments. His vision stretched to the idea that people would want to work on parallel documents, side by side on a screen.

The hope of Xanadu

For years, Nelson has attempted to make his ideas come to fruition. The largest concentration of his effort has been to develop a unified structure for computer documents that will handle this massive degree of interconnection and cross-referencing. The project has worked under the trademark of Xanadu and has had many setbacks, while at the same time he has seen the Web come into existence.

Far from being excited by the Web, Nelson sees it as being little better than presenting all the restrictions of paper on computer screens. For him, the Web is still constraining the way that people think, and not providing a tool that will fully liberate the human mind.

Xanadu may never get off the ground, but in coining the term "hypertext" Nelson has affected our ways of thinking, by creating the concepts of multiple interconnected texts and non-linear approaches to reading documents.

Timeline

1937
Born in Chicago, Illinois, June 17. His father, Ralph Nelson, is a film director, his mother, Celeste Holm, an actress
1955–1959
Studies philosophy at Swarthmore College
1959–1960
Enters graduate school at the University of Chicago
1960–1963
Moves to Harvard to take a Master's degree in sociology
1965
Coins the terms "hypertext" and "hypermedia"
1974
Self-publishes his book *Computer Lib/Dream Machines*, which foretells a world of personal computing
1979
Nelson spends an exciting summer with a team of "followers" who form XOC, Inc., and work on his Xanadu concept
1987
Apple introduces the Hypercard program, which allows users to construct webs of links within their personal computers
1988–1992
Xanadu project is supported by Autodesk, Inc., under the control of the XOC group with Nelson acting as a spokesman. They do not deliver a working system
1994–
Takes appointments in Japan, firstly at Hyperlab in Sapporo, and then as visiting professor of environmental information at Keio University, Fujisawa

Computers operate by handling strings of 0s and 1s. These can be easily converted into electronic pulses and shot along telecommunication lines. Over the 1970s a grid of inter-linked computers were organized to form an Internet, and people with computing know-how could use it to transmit documents. Then in 1989, engineer Tim Berners-Lee created the World Wide Web, making it so easy to use the Internet that anyone with a personal computer and a telephone line could join in. The concept has led to one of the most remarkable communication revolutions that has ever occurred, and in so doing has altered the ways that people interact.

1955–

Acquaintances and Influences
— Doug Englbart (1925–)
— Ted Nelson (1937–)
— Robert Khan (1938–)
— Vinton Cerf (1943–)
— Michael Dertouzos (1936–2001)

Tim Berners-Lee

Adding the Web to the Net

The history of communication has been marked by a few innovations that have been spread over great stretches of time. The ancient Egyptians invented papyrus some 30 centuries before Christ, and then in 105 A.D. Ts'ai Lun, a member of the Imperial court of Peking, invented paper. It took a thousand years before Pi Cheng, in the years between 1041 and 1048, invented a machine that could use movable characters to print onto the paper and another five hundred years before Johannes Gutenberg and William Caxton developed printing presses that gave birth to the European printing industry. Arguably a further five hundred years passed before the next significant change occurred.

In 1980, Tim Berners-Lee was working on a six-month contract as a software engineer at CERN, the European Laboratory for Particle Physics, in Geneva. He was frustrated by the difficulties that he always had organizing his notes, and was keen to see if there was a better way of allowing the computer to give access to his stored information. He wanted to create a system that would start to resemble the brain in the way that it keeps track of all the connections and links between different pieces of data.

The result was a piece of computer software that he called Enquire, after a Victorian encyclopedia called *Enquire Within Upon Everything*. Enquire enabled Berners-Lee to create "hypertext" documents — documents that had links to other files within his computer. It was a step forward, but it was limited. If the file was on someone else's computer then there was no easy way of making that connection.

At the same time Vinton Cerf and Robert Khan were busy creating the Internet. This set of inter-linked computers provided a global network of technology that enabled information to be broken into small packets and posted to specific addresses. The problem with the Internet was that it was relatively laborious to send and receive documents. Berners-Lee was convinced that it could be employed more powerfully to give people instant access to information and files held on each other's computers. He developed a simple coding system that he branded HTML or HyperText Mark-up Language, and designed a system of addresses that gave each page of information a unique name — a URL or Universal Resource Locator.

On top of this he created a quick set of rules that he called HyperText Transfer Protocol (HTTP), which allowed the documents to be linked even if they were present on different computers. His final act was to put together a simple browser program that would allow people to see documents on their computer screens.

In 1991 he unveiled the results — what is now commonly known as the World Wide Web.

Impact on life

The Internet is therefore a physical structure of computers, cables, and conventions for sending packets of data between computers. The World Wide Web, however, is much harder to pin down. It is in effect an abstract space of information that exists on the Net. The Web has made the power of the Internet readily available. Part of its beauty is in its universality — a single hyperlink code can point to data in a user's own computer, or to data stored on any other computer that is connected to the Internet.

Just as the printed book enabled many more people to join in debates and become educated, this new mode of communication has had a number of intriguing consequences. To start with, the Web can create social structures that are independent of geography. Most of the time when a person is looking at a web site it is unclear where the author of the site actually exists. This has given rise to unprecedented opportunities for new freedoms of expression, but has also become a censor's nightmare, as value and vice exist side by side.

Berners-Lee's dream was for a common space that encourages people to share information, where a simple hypertext code can connect any two addresses. He wanted something that would have no central organization, no central database or controller. He also hoped that it would become absorbed into people's everyday lives, not only providing data, but also enabling computers to help people analyze the ways that we can collaborate more effectively.

From an initial start in 1991, the World Wide Web has seen colossal growth. In itself, this has altered the world, but Berners-Lee is one of the first to admit that it is difficult to prophesy its ultimate impact.

Timeline

1955
Born in London, England, on June 8

1976
Graduates from Queen's College, Oxford, with a degree in physics

1980
While working on a six-month contract at CERN, the particle physics laboratory in Geneva, Switzerland, Berners-Lee writes Enquire, a computer program that formed the foundation of the World Wide Web

1984
Back at CERN, he works on ways of letting scientists gain instant access to data stored on computers

1989
Suggests the possibility of a global hypertext project which he calls the World Wide Web

1991
Berners-Lee writes some news articles suggesting that people download server and browser software, a move that launches the Web on the Internet

1994
Joins the Laboratory for Computer Science at Massachusetts Institute of Technology (MIT) and becomes director of the World Wide Web Consortium, an open forum of companies and organizations that coordinate Web development world-wide

1999
Becomes the first holder of the 3Com Founders chair at MIT

The world within us

As the 18th century closed, a few people were starting to question the Judaeo-Christian world view that the Earth and all its creatures had been created at about 4000 B.C. As the new century opened, the likes of Erasmus Darwin and Jean-Baptiste Lamarck began to propose another view. For them, simple life forms had occurred spontaneously and been driven to become increasingly complex by a "vital force." Then came Erasmus's grandson Charles Darwin, who simultaneously with Alfred Wallace deduced that species could adapt, and those that were better suited to their environments would have most chance of survival. Science and society encountered evolution.

Charles Darwin 1809–1882	Acquaintances	
	— Erasmus Darwin (1731–1802)	— Robert FitzRoy (1805–1865)
	— Thomas Robert Malthus (1766–1834)	— Samuel Wilberforce (1805–1873)
Alfred Wallace 1823–1913	— Adam Sedgwick (1785–1873)	— Joseph Dalton Hooker (1817–1911)
	— John Henslow (1796–1861)	— Herbert Spencer (1820–1903)
	— Charles Lyell (1797–1875)	— Thomas Huxley (1825–1895)
	— Richard Owen (1804–1892)	

Charles Darwin & Alfred Wallace

The voyage of HMS *Beagle*

Charles Darwin was born into a world that basically accepted that fish had fins because a creator-God had wanted them to swim, and that cheetahs could run fast because the same creator wanted them to catch food. It was a view that Darwin seems initially to have taken at face value, but was already questioning by the time he began to study divinity at Cambridge. He had previously dropped out of studying medicine because he found it difficult to watch operations on people in the days before anesthetics had been invented.

While at Cambridge he went on field trips with friends looking at geological formations and their fossils. He and others started to question whether, for example, birds had wings because the creator wanted them to fly, or whether birds could fly because they had wings.

At Cambridge, Darwin had to read two books by Archbishop William Paley: *View of the Evidences of Christianity* (1794) and *Natural Theology* (1802). The first argued that you could build up a thesis of events such as the resurrection of Christ by studying indirect evidence; the second reasoned that the natural world bears all the hallmarks of a creator. Both influenced his thinking, though not necessarily as the author had intended. *Evidences* taught him a scheme for seeking evidence of unrepeatable historic events, and *Natural Theology* posed many questions that set his mind racing — questions such as how could an eye originate by chance?

So, searching for evidence of the basic origins of life, Darwin set sail on HMS *Beagle* — he was rich enough to pay his own way, as his mother was a daughter of the wealthy maker of fine porcelain Josiah Wedgwood. His father was not keen on the idea, but eventually allowed him to travel as the companion of the ship's captain, Robert FitzRoy.

On his voyage, Darwin read Charles Lyell's *Principles of Geology*, which argued that the world was being shaped and reshaped by constantly acting geological forces. Known as the "steady state" view, it suggested that the world was the product of steady change, rather than of sudden catastrophic events such as a few days of creation and the odd flood.

When he arrived in South America, Darwin found more evidence supporting Lyell's steady state theory. He started building on this with observations from fossil evidence that seemed to indicate some form of progression from simple to complex life forms. Between September 1832 and August 1833, while staying at Bahia Blanca just south of Buenos Aires on the Atlantic coast of South America, he dug up bones of extinct giant species of animals that resembled present-day sloths, armadillos, and llamas.

By 1835 they had reached South America's west coast, and on February 20 Darwin experienced an earthquake that destroyed the town of Concepcion in seconds as the land was lifted by between one and three yards. He also found fossilized seashells 4,000 yards up in the mountains, and started to speculate that these had probably got there because many such earthquakes had driven the ground higher and higher through thousands of years.

above *The HMS* Beagle, *the ship that carried Charles Darwin during the voyage that inspired his theory of evolution. Darwin joined the crew in December 1831 as an unpaid naturalist for a five-year journey.*

opposite *Charles Darwin (left) and Alfred Wallace (right).*

His most famous two months were spent on the Galapagos Islands in the Pacific Ocean, from September to October in 1835. This equatorial group of islands had been created by volcanoes, and as far as Darwin could see the islands were not particularly old. Certainly not in comparison with the continents. He was, therefore, surprised by the diversity of life that he found on the islands. He found 71 species of plant on James Island, of which 30 were found nowhere else in the world, and a further eight were found only on other islands in this little group. He was intrigued by the variety of species of finches, each with differently shaped beaks that seemed well suited to the particular food available on each island.

So on October 2, 1836, Darwin returned to Falmouth with trunk-loads of specimens and notes, and a packed mind. Now he needed to sit down and try to make sense of the data.

Engaging with Wallace

Unknown to Darwin, someone else was developing similar ideas. Having had an unfortunate start to life, which included having to abandon his education at Hertford Grammar School when his father ran out of money, Alfred Wallace formed an expedition to South America in 1848. Typical of his bad luck, his ship caught fire and sank on the return voyage and he was fortunate to be rescued 10 days later. In March 1854 he set off again, this time to the Far East. He was a prodigious collector, and amassed over 125,000 specimens. In February 1855 Wallace gathered some of his ideas together in an essay called "On the Law which has Regulated the Introduction of New Species." The paper sets out the principles of evolution.

Lyell read the paper and showed it to Darwin, who, it appears, was unimpressed, though the seed of an idea was sown. By this point, Darwin was living in Kent, and suffering from a lack of energy, probably due to a disease that he had picked up in South America. He was sorting through his data, but Lyell persuaded him to get his ideas down on paper—to write a book. Darwin was meticulous and not about to be pushed. After all, the ideas that he was framing were revolutionary, and he seems to have been anxious not to make a fool of himself by presenting an unsupported theory.

The stimulus to announce his thoughts came from a letter he received on June 18, 1858. The letter was from Wallace, and contained an essay: "On the Tendency of Varieties to Depart Indefinitely from the Original Type." Wallace had been recovering from malaria, when he remembered something he had read a few years earlier. Thomas Malthus had suggested that the size of populations, and in particular human populations, was limited by the food supply. The problem was that, as far as Malthus was concerned, populations always grew faster than their ability to produce food, and consequently there would always be hunger and famine.

Wallace suddenly realized that if this was so, the individuals most suited to a particular situation would survive, while the weaker ones would die. It was a case of "the survival of the fittest." And the fittest would pass on their good features to future generations.

When Darwin showed the letter to Lyell and a few other associates they were convinced—no time should be lost.

They took the letter and Darwin's work and presented it at the next meeting of the Linnean Society. This occurred on July 1, 1858, and just over a year later Darwin's book came out in print. Evolution was now up for debate. Darwin was well on the way to being the first person with a book on the subject, he had good financial resources and friends in high places—and thus soon became known as the author, not only of the book, but also of the theory.

Huxley and the bishop

Any review of the history of science soon finds that its discoveries have implications that run far beyond the laboratory. Sometimes the findings make rulers uncomfortable; at other times they cause religious leaders to question long-held assumptions. *The Origin of Species* is certainly no exception to this rule. With the best part of 150 years since its publication, it is now difficult to be sure exactly how much of an argument it created between science and the Church: the story has been told so many times by people trying to promote one view or another, that the truth of the clash is difficult to reveal.

below *Darwin found giant tortoises on four of the Galapagos islands, but on each island the animals had distinctive shells. It was one of a series of observations that led him to think of evolution.*

Certainly Darwin's theory could only operate within a framework where the world was many, many thousands of years old. This conflicted with a literal reading of the Bible, which suggests that the world may be as young as 6,000 years. It also challenged head-on the concept that all of the universe and all living things were created in six days, after which little has changed.

One of the more notorious episodes in the debate occurred at a meeting of the British Association in 1860, when Thomas Huxley debated the theory of evolution with Bishop Samuel Wilberforce. At the time, Huxley struggled to answer many of the bishop's probing questions and at one point, in apparent self-defense, said that he would rather have a miserable ape for a grandfather than a bishop.

It appears that the bishop challenged the theory not on the basis of theology, but because of the lack of critical supporting evidence. Thankfully for Huxley, botanist Joseph Hooker was also present, and managed to give some form of answer to the majority of the bishop's points. At the time of the debate, the bishop was seen by most people as being on the winning side, and only with the arrival of new data has general opinion swung in favor of Darwin's theories.

While some people still see a conflict between science and religion, most realize that science looks at the mechanisms and asks how things occur, while religion looks at the meanings and ask why they happen. Consequently, the two can interact, but because they are asking different questions, it is incorrect to suggest that there is genuine conflict.

Timeline – Charles Darwin

1809
Born at Shrewsbury, England, on February 9
1825–1827
Studies medicine at the University of Edinburgh, and also goes on geological trips with Robert Jameson
1827–1831
Studies divinity at Christ's College, Cambridge, and also studies botany with John Henslow and geology with Adam Sedgwick
1831
On December 27, Darwin sets sail on HMS *Beagle* as an unpaid naturalist. They return on October 2, 1836
1838
Becomes secretary of the Geological Society
1839
Is elected a fellow of the Royal Society and marries his cousin, Emma Wedgwood, the daughter of the pottery magnate. The couple subsequently have four sons
1842
The Darwins move to Down House, Kent
Darwin starts work on the first draft of his transmutation theory, which in 1844 becomes a 230-page essay
1856
Charles Lyell advises Darwin to publish his ideas. Darwin starts work on his "big book"
1858
On June 18 Darwin receives a letter from Alfred Wallace, setting out his view on the evolution of species, and on July 1, Darwin presents their ideas before the Linnean Society in London in Wallace's absence
1859
Publishes *The Origin of Species by Means of Natural Selection — or the preservation of favored races in the struggle for life*
1871
Publishes *Descent of Man*
1882
Dies at Down House on April 19

Timeline – Alfred Wallace

1823
Born at Usk, Monmouthshire, on January 8
1848–1852
Goes on an expedition to South America with Bates
1854
Leaves England for the Far East
1855
While in Malaysia, Wallace publishes an essay in an English language journal giving his view of the way that species develop
1858
Without his knowing, parts of his essays are read at the Linnean Society in London
1862
Returns to England
1864
Presents "The Origin of Human Races Deduced From The Theory of 'Natural Selection'" to the Anthropological Society in London
1893
Elected a fellow of the Royal Society
1913
Dies at Old Orchard, near Wimborne, Dorset, England

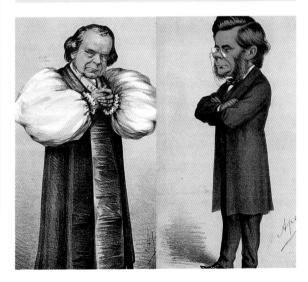

right *Bishop Samuel Wilberforce and Thomas Huxley's famous debate in Oxford is often represented as the moment when science clashed with religion.*

In the mid 19th century scientists were aware that progeny inherited characteristics from their parents, but no one had suggested a plausible mechanism. Austrian monk Johann Gregor Mendel's now famous work on peas indicated that features of the plant were passed from one generation to the next by some physical "element." He realized that each characteristic of a plant was inherited independently, and that the ratios of plants inheriting each trait could be statistically predicted. The work effectively lay undiscovered until 16 years after his death, when other scientists realized that it provided a foundation to the newly emerging science of genetics.

1822–1884

Acquaintances
—F.C. Napp (1792–1867)
—Carl Wilhelm von Nägeli (1817–1891)

Gregor Mendel

Hybrids and inheritance

When Johann Mendel set out to find rules that would explain the way that plants and animals inherited different traits, the general feeling was that organisms adapt to their surroundings during their lives and pass on any beneficial lessons to their progeny. The transmission of hereditary characteristics from parents to offspring was just one part of the evolutionary adaptation, and possibly not even the most important part.

People were aware that offspring, be they plants or animals, consisted of mixtures of the physical characteristics seen in their parents—a situation made obvious by parents looking at their baby and saying, "She's got her mother's nose." They realized that individuals are, to an extent, mixtures of their parents.

In his first paper, *Versuche über Pflanzen-Hybriden*, published in 1866, Mendel wrote that he wanted to discover a "generally applicable law of the formation and development of hybrids." To address this, he carried out carefully regulated breeding experiments with a strain of pea, *Pisum sativum*. He chose this plant because it was cheap, and it came in many strains that when bred in isolation produced identical offspring, but which gave more complex results when mixed to produce hybrids. In one experiment he crossed one strain that produced smooth, round peas with another that produced wrinkled peas. The offspring all had round seed. When these offspring were crossed with themselves, some of the progeny had round and some had wrinkled seed. Intriguingly the ratio was almost perfectly three round peas for every one wrinkled seed.

He repeated the experiment, looking at the way different colors were passed from one generation to the next and found that the same rule applied. He also noted that the inheritance of color occurred independently of the inheritance of pea shape.

Performing the experiments required diligence, but making sense of them needed genius. Mendel realized that his results could only occur if characteristics were carried from one generation to another by some material means. He didn't speculate whether this was fluid, gas, or some solid particle, but did insist that it was an "element" and not a mystic "potential" as others were suggesting. He also showed that there were sets of these "elements" for each characteristic, for example a smooth-pea element and a wrinkled-pea element and that contrary to popular belief these could not merge. Any individual pea always had one or the other, and never a mixture of the two.

He was also convinced that the elements regulating different features acted independently of each other, so the inheritance of one particular color of flower was not influenced by the inheritance of pea-shape.

Added together, this was new and radical. It made inheritance a matter of physical elements and statistics, and removed any mystic or spiritual involvement. In a letter to Carl Wilhelm von Nägeli dated April 18, 1867, Mendel reveals his own impression of the impact of his findings, and the consequent need to ensure that his work was robust. "I knew that the results I obtained were not easily compatible with our contemporary scientific knowledge, and that under the circumstances publication of one such isolated experiment was doubly dangerous; dangerous for the experimenter and for the cause he represented." To prove his point he studied some 28,000 pea plants and wrote just two scientific papers.

A nervous pioneer

Mendel was born on July 22, 1822, in Heizendort. At that point in history, this village was in the Silesia region of Austria, but as boundaries have moved it is now called Hyncice and is part of the Czech Republic. He was the second child of Anton and Rosine Mendel, who, although they were peasant farmers, struggled to put Mendel through school, in the hope that he would have a better life.

While still at school his father had an accident and Mendel tried to earn a living by teaching. The stress of this appears to have been too much and his health broke down. A similar breakdown occurred later when in 1850 he took the examination to become a teacher. His professor was, however, sufficiently impressed with Mendel's work to recommend that Mendel join the Augustinian monastery at Brno.

His abbot, F.C. Napp, sent him to the University of Vienna where between 1851 and 1853 he studied physics, chemistry, mathematics, zoology, and botany, but he still failed to get a teaching certificate. Napp supported teaching in agriculture and was an active member of the local sheep breeders' association. He encouraged Mendel to attend the Brno Natural History Society, which at that time frequently debated the theory of natural selection.

When Mendel died in Brno on January 6, 1884, he was convinced he had made an important discovery and was disappointed that no one cared. It was only in 1900 that three other scientists, Dutchman Hugo de Vries, German Carl Correns, and Austrian Erik Tschermak, who had each independently repeated his work, discovered similar rules and then found that Mendel had got there before them.

above *The monastic garden at Brno Monastery where Mendel grew his plants and brought genetics to the world.*

By the time that Barbara McClintock entered research, the concepts of genetics were becoming clear. Genetic information was stored in the center of cells in structures called chromosomes. While the way that these rod-like structures were made was far from understood, geneticists could start to map out where particular bits of information were stored on each one. Having a fixed location was obviously important so that the cell could organize its library of information. Then McClintock announced that some genes could jump from place to place within the chromosomes. The simple concept of a static library was suddenly replaced by a more complex reality.

Barbara McClintock

1902–1992

Acquaintances
—Rollins Emerson (1873–1947)
—Lewis Stadler (1896–1954)
—George Wells Beadle (1903–1989)
—Marcus M. Rhoades (1903–1991)
—Harriet Creighton (1909–)

Studying chromosomes

When Barbara McClintock graduated from Cornell University in 1923, scientists knew that chromosomes were the structures inside cells that contained the heritable "elements" that Gregor Mendel had identified in the previous century. Chromosomes were intriguing. These stripy rod-like features could be seen under a microscope in the center of cells that were in the process of dividing. At this point the chromosomes produced copies of themselves, so that each new cell received a complete set.

Chromosomes were therefore likened to reproducible banks of data, or libraries. People were beginning to talk about individual characteristics being caused by the information stored in chunks of the chromosome, and calling these chunks "genes." This gave vital clues about how information could be passed from one generation to another, but said little about the intricate working of these remarkable organelles.

When offered the opportunity of studying for a Ph.D. in genetics, McClintock was thrilled. She had already decided that she wanted to discover how the information was stored on these chromosomes, and how the cell made use of that information. Shortly after receiving her Ph.D. in 1927, McClintock began work with fellow geneticist Harriet Creighton. Together they showed that chromosomes were not fixed entities. By now scientists had realized that chromosomes occur in pairs. The two women found that during cell division each pair of chromosomes came together and exchanged stretches of their material. This meant that newly formed cells had combinations of genetic instructions that had never been seen before. A bit later, McClintock discovered that if a chromosome was broken by, for example, exposing the cell to radiation, then the fragments would form circles. She decided that healthy chromosomes normally didn't do this because of special features at their tips—features she called telomeres.

Jumping genes

Then in the summer of 1944 McClintock started to get some puzzling results. Each cell in a maize plant has 10 pairs of chromosomes, and she was beginning to form a good idea of the location of particular genes on individual chromosomes. However, over the next five years she collected data suggesting that certain fragments of the chromosome could quite simply jump from one place to another. The idea was so revolutionary that when, in 1956, she first presented her theory most scientists completely ignored it.

Undaunted by the apathy from her colleagues, McClintock pursued her ideas. The first mobile element that she identified existed on chromosome number 9. This particular jumping gene caused the chromosome to break wherever it inserted itself, preventing the nearby genes from working.

With further research she found that jumping genes could move only to certain locations. The reason for this appeared to be that they were involved in deliberately inactivating particular genes. In fact they formed part of the system for controlling the way that a cell is working.

Later on, McClintock found that mobile elements could also move from one chromosome to another, enabling a gene on one chromosome to influence the activity of a gene on another chromosome. This revealed another layer of complexity in the way that genes are controlled.

While the discovery of "jumping genes" profoundly influenced our understanding of genetics, most of the time McClintock worked on her own. It is interesting that like Mendel before her, her work only gained the recognition it deserved many years after she announced her findings.

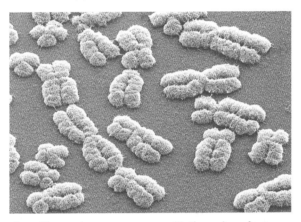

above *Human chromosomes, present in the nucleus of every cell of the body. Most human cells contain 46 chromosomes.*

Timeline

1902
Born on June 16, in Hartford, Connecticut, McClintock is brought up in Brooklyn, New York
1921
At Cornell University she attends the only course in genetics open to undergraduate students. She graduates in 1923
1927
Receives a Ph.D. in botany from Cornell University
1936
Moves to the University of Missouri to work with Lewis Stadler
1941
Spends the summer at Cold Spring Harbor, as the guest investigator of maize geneticist Marcus Rhoades, and never leaves
1944
Elected a member of the National Academy of Sciences—only the third woman to be given this honor. In this year she identifies the presence of "jumping genes"
1950
Publishes data about "jumping genes"
1951
Her work is not widely accepted when first presented at a Cold Spring Harbor Symposium
1983
Receives the Nobel Prize for Physiology or Medicine
1992
Dies on September 2 at Huntington, New York

Frederick Sanger

1918–

Acquaintances
— Max Ferdinand Perutz (1914–2002)
— John Cowdery Kendrew (1917–1997)
— Paul Berg (1926–)
— Aaron Klug (1926–)
— Walter Gilbert (1932–)

Proteins are fundamental to life. They act as building blocks within cells, as messengers both within and between cells, and as enzymes that encourage specific chemical reactions to take place. When Frederick Sanger started his work, scientists knew that proteins were built of amino acids. His first outstanding piece of work was to determine how these amino acids linked to form the small but important protein, insulin. His second critical contribution to the world was to develop a method of finding the code of the genetic instructions that enable cells to build proteins in the first place. In recognition he received two Nobel Prizes.

Unpacking protein

Break any living organism down and you will be left with a collection of fats, carbohydrates, minerals, water, and protein. Of these, the protein is the most complex element, and biochemists now know that its precise structure determines the way that it functions within an organism. When Frederick Sanger completed his Ph.D. in 1943, this was not yet realized.

Sanger started to take a close look at insulin. Around the start of the 20th century scientists found evidence linking the disease of diabetes with the lack of a chemical component produced in the pancreas. Further research narrowed the site of production to specialist cells called "islets of Langerhans." This chemical became known as insulin, and in 1922 Canadian biochemist Bertram Collip gave an extract of cow pancreas to 14-year-old Leonard Thompson, a teenager with diabetes. The injection radically improved his health. But still no one knew what insulin was, other than that it appeared to be some form of protein.

At that stage, the best you could do was look at the protein using an electron microscope, but the images revealed little of protein's fine structure. Herman Emil Fischer had won a Nobel Prize in 1902 for discovering that protein molecules contained long strings of so-called amino acids and that you could break down protein into a soup of amino acids by boiling it with strong acids. Scientists had also come to the somewhat startling conclusion that there were only 22 different types of amino acids occurring in nature—all of the amazing variety of proteins and life forms were built of just 22 different building blocks.

To start with, Sanger developed a way of marking the end of the protein chain by adding a dye, dinitrofluorbenzene. This locks to the last amino acid so strongly that it remains there even if the chain is broken into its individual amino acids. His first discovery showed that the task would be slightly easier than he had anticipated. Insulin was made of two different chains, one consisting of 20 and the other 31 amino acids. This meant that he could study each individually, substantially reducing the level of complexity.

He developed ways of breaking the chain into lengths of two, three, four, five, or more amino acids, and could determine the sequence in each short section. Then it was a matter of sorting out a total sequence that could make sense of all the fragments. The task took eight years, but he got there. He knew the sequence of amino acids that made up insulin's two chains.

That information on its own would have had little long-term value, but now there was a way of finding out the composition of every protein in living organisms. Once the sequence of amino acids was known, protein chemists could start to speculate how the chain wrapped around itself to form the physical shapes in the molecule that allow it to carry out its biological functions. There was also a good chance that people could create the protein artificially and start to manufacture life-saving drugs.

And for my next trick

Speaking in Stockholm at the Nobel ceremony, Professor A. Tiselius, a member of the Nobel committee for Chemistry, said prophetically: "It was Alfred Nobel's intention that his prizes should not be considered as awards for achievement done, but that they should serve as encouragement for future work. We are confident that you are a worthy recipient of the Nobel award also in this sense."

Indeed Sanger claims that the award did stimulate his career, as the recognition of his work gave him renewed confidence, and the opportunity to get better research facilities and attract like-minded colleagues. Consequently, in 1980 Sanger was back in Stockholm receiving another Prize — one of only four people to pick up two of these prestigious medals. Once again the award recognized his

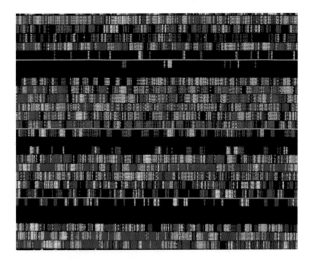

above *Human DNA sequence. DNA consists of two long strands linked by the interactions of bases along their lengths. Each color represents a specific base. The sequence of bases makes up the genetic code in the form of genes, segments of DNA which have specific functions within an organism.*

development of a method of determining biological sequences, but this time it was the sequence of genetic code "characters" that are stored on chromosomes.

Unravelling DNA

Between his first and second appearance on the Stockholm stage scientists had discovered the structural basis of chromosomes. They were made of vast chains of so-called deoxyribonucleic acid, or DNA. Each link was a nucleotide, with one of just four different "bases" attached—adenine, guanine, thymine, and cytosine, normally abbreviated to A, G, T, and C. The assumption was that the exact sequence of these bases formed a genetic language, a biological code that held the instructions enabling a cell to build all the different proteins that it needed.

The question was, what was the code? What was the order of these bases? This was going to be no simple task. Sanger had spent eight years working out a way of sorting out the order of 51 amino acids in a molecule of insulin, but that now looked like child's play. The first set of genetic code that he addressed was the chromosome in phi-X174, a virus that attacks bacteria. This is a small chromosome, but even so consists of 5,375 nucleotides.

His technique was reminiscent of the way that he revealed the sequence in proteins, and in recent years the process of analyzing the DNA sequences has been

automated. Now we can start to look at the biological mechanism that enables cells to build protein, and start to ask questions about the underlying reasons for differences between individuals. This can be particularly important as some of those differences lead to genetic disease.

One consequence of automation is that scientists could tackle large projects such as finding the code-sequence carried in the DNA of human beings, a sequence that contains 3 billion "letters." A landmark was reached in this quest in 2000, when scientists announced that they had completed a "draft" version of this code, and expected to have an almost completely accurate listing within the following five or so years.

Shortly after receiving his second Nobel medal, Sanger retired, but his name lives on in science, not least because one of the world's leading centers of genetic research—the Wellcome Trust Sanger Institute, Cambridge, England—bears his name.

Timeline

1918
Born on August 13, at Rendcombe, Gloucestershire
1939
Receives a degree in natural sciences at St. John's College, Cambridge, England
1940
Starts work in the department of biochemistry at Cambridge, and marries Margaret Joan Howe. The couple subsequently have three children Robin, Peter, and Sally Joan
1940–1943
As a conscientious objector through World War II he studies the amino acid lysine and problems concerning nitrogen in potatoes
1943
Obtains his Ph.D.
1951
Becomes a salaried member of staff of the Medical Research Council
1958
Receives his first Nobel Prize for discovering the sequence of amino acids in the protein insulin
1962
Moves to a newly built laboratory of molecular biology in Cambridge, along with the likes of Max Perutz, Francis Crick, John Kendrew, and Aaron Klug
1980
Receives his second Nobel Prize, this time for devising a method of finding the sequence of bases in DNA
2000
Sees the publication of the draft sequence of the three billion bases that make up the human genome

above *The insulin hormone molecule is made up of two chains of amino acids: the A chain (yellow and green), and the B chain (blue). These chains are held together by two sulphur bonds with the zinc atoms (white) at the center.*

Arguably the single most profound discovery of the 20th century was that of the structure of DNA, the material that stores genetic information in chromosomes. By the time that James Watson and Francis Crick had shown that it was a "double helix," geneticists knew that it stored all the information needed to build and run an organism. While the structure was interesting in itself, the real value of the discovery was that the mechanism allowing chromosomes to both store information and, more importantly, to copy themselves, suddenly became clear. The double helix now stands as an icon of the scientific understanding of life.

James Watson	Acquaintances
1928–	—Hermann Joseph Müller (1890–1967)
	—Salvador Edward Luria (1912–1991)
Francis Crick	—Max Ferdinand Perutz (1914–2002)
1916–	—Maurice Hugh Frederick Wilkins (1916–)
	—John Cowdery Kendrew (1917–1997)

James Watson
& Francis Crick

Helical storage

Critical to biology is the idea that a living organism can reproduce — it can generate a new member of its species. Equally critical, for all but one-cell organisms like bacteria, is the organism's need to grow. Both reproduction and growth demand that an organism forms new cells, and that all of the information needed for its existence is passed into all of the new cells.

Gregor Mendel showed that the information had some physical nature, and in 1928 British medical officer Frederick Griffith discovered that genetic information could be passed from heat-killed bacteria to living bacteria. Quite clearly this genetic element could withstand heat. Sixteen years later, Canadian-born bacteriologist Oswald Avery claimed that this heat-stable element was a vast molecule of deoxyribonucleic acid — more commonly referred to as DNA. At first no one took their comments seriously. After all, everything that was known about the molecule up to that point suggested that it was far too simple to perform such a complex function as storing vast quantities of data. In addition, scientists were well aware

that DNA was made of deoxyribose sugars, phosphate molecules, and just four different types of "bases" — adenine (A), guanine (G), cytosine (C), and thymine (T).

People started taking DNA more seriously after biochemist Erwin Chargaff discovered that the composition of DNA was the same in cells that come from one species, but radically different between species. He was also puzzled to discover that cells' DNA always had equal quantities of A and T and equal quantities of C and G. The race was on to discover the nature of this molecule and find out how it stored information in a way that could easily be copied each time a new cell is made.

In 1951 a 31-year-old X-ray crystallographer gave a lecture suggesting that DNA could be a big helix. Her name was Rosalind Franklin, and she worked in King's College, London. Her comment came from ideas generated by looking at X-ray-generated photos taken of crystals of DNA. She was convinced that the sugar and phosphate units formed a backbone to DNA that lay on the outer facet of the molecule. Present at the lecture was American geneticist James Watson, who had recently arrived in England and was working with Francis Crick in Cambridge. The pair were fighting to make sense of DNA.

By this time many scientists were convinced that DNA was a chain that wrapped around itself in some form of spiral. For example, American chemist Linus Pauling had suggested that it had three intertwining strands. Watson and Crick were working together in their laboratory with models of all of the components of DNA, trying to see if they could find a way of constructing the molecule that made sense of the data.

Franklin, on the other hand, was working much more on her own. Life was not easy for a woman in science, and one of Franklin's colleagues, Maurice Wilkins (1916–), seems to have been less than ready to accept that she was a peer. However, he did recognize the value of her work and made copies of her X-ray photos, copies he showed one day to Watson. When Watson saw Franklin's photographs he claims that he became confident — DNA must be a helix.

Back in the Cavendish Laboratory, Watson and Crick concentrated on building a helix, and by March 7, 1953, they had a solution. DNA was a double helix, with two strands twined around each other. It fitted with Franklin's X-ray data. In addition, their model explained Chargaff's data, as the strands were linked together by the bases: A always linked to a T, and C always held on to a G. What's more, this model suggested a mechanism by which the molecule could make copies of itself. The two strands could peel apart and act as templates for new strands to build on. The concept was shocking in its simplicity.

Learning a language

Realizing that DNAcontained long strands of sugar molecules linked by these four different bases was one thing. Working out how these bases stored information was quite another. In his Nobel Lecture given on December 11, 1962, Crick spelt out the problem. DNA gave the instructions to connect the 20 or so different amino acids

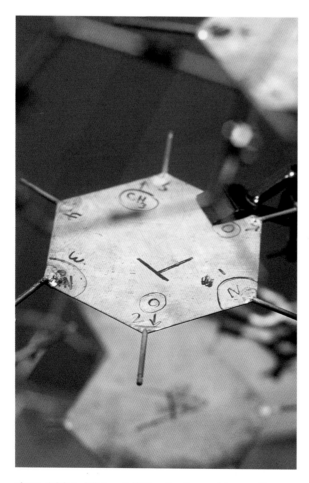

above *Crick and Watson's DNA molecular model, 1953. The metal plates represent the four bases whose complementary arrangement immediately suggested a possible copying mechanism for the genetic material.*

together to build proteins. But DNA had only four different bases. Clearly these bases must be grouped so that a number of bases related to each amino acid. The first question was, how many?

Crick pointed out that if the "code" was written in pairs of bases, using all the combinations would only give 16 options. This was clearly not enough. Using a three-letter code would generate 64 possibilities, which represented a considerable excess. It turned out that three was the right number. Every three bases in the DNA gives the instruction for one specific amino acid. It is as if genetics uses a four-letter language and three-letter words.

What of Franklin?

Tragically, Rosalind Franklin died at the age of 37 due to cancer, and since her death there has been considerable controversy over the way that history has recorded her contribution to the DNA story. When she was working in London it was hard for a woman to contribute fully as, for example, only males were allowed in the university dining rooms and after-hours Franklin's colleagues socialized in male-only pubs. Consequently she was excluded from much of the scientific gossip and banter that plays such a critical part in grappling with ideas.

Her untimely death also meant that the Nobel Committee was unable to salute her involvement in the discovery. Instead they gave the award jointly to Crick, Watson, and Wilkins, where, intriguingly, Wilkins was largely credited for creating the vital X-ray photographs.

So much for history. When Professor A. Engström presented the three men to the Swedish king so that they could receive their Nobel medals, he said, "Today no one can really ascertain the consequences of this new exact knowledge of the mechanism of heredity. We can foresee new possibilities to conquer disease and to gain better knowledge of the interaction of heredity and environment and a greater understanding for the mechanisms of the origin of life."

Since then, biological science has largely concentrated on trying to turn these possibilities into reality.

Timeline – Francis Crick

1916
Born on June 8, in Northampton, England
1937
Receives a degree in physics at University College London
1939
Works for the British admiralty during World War II, mainly in connection with acoustic and magnetic mines
1940
Marries Ruth Doreen Dodd, but divorced in 1947. The couple had a son, Michael
1947
Marries Odile Speed, and the couple have two daughters, Gabrielle and Jacqueline
1949
Joins the Medical Research Council research group in Cambridge, headed by Max Perutz
1951
Starts collaborating with Watson
1953
Reveals the structure of DNA
1954
Receives a Ph.D. His thesis was entitled "X-ray Diffraction: Polypeptides and Proteins"
1959
Is elected a fellow of the Royal Society
1961
Crick's team finds the genetic code for proteins
1962
Shares the Nobel Prize with Watson and Maurice Wilkins
1977
Begins brain research at the Salk Institute, California
1928
Born on April 6, in Chicago, USA

Timeline – James Watson

1947
Receives a degree in zoology at the University of Chicago
1950
Receives Ph.D. from Indiana University in Bloomington
1951
Starts collaborating with Crick
1953
Reveals the structure of DNA
1953–1955
Works at the California Institute of Technology
1955–1956
Returns for a short spell of work at the Cavendish Laboratories in Cambridge
1956
Joins the department of biology at Harvard
1962
Shares the Nobel Prize with Crick and Maurice Wilkins
1968
Publishes his account of the discovery in *The Double Helix*
Becomes director of Cold Spring Harbor Laboratory
1988
Is named as head of the U.S. Human Genome Project, though resigns before its completion

William Harvey

1578–1657

Acquaintances
- Hieronymus Fabricus of Aquapendente (1537–1619)
- Francis Bacon (1561–1626)
- Galileo Galilei (1564–1642)
- King James I (1566–1626)
- Gaspare Aselli (1581–1626)
- René Descartes (1596–1650)
- King Charles I (1600–1649)
- Robert Boyle (1627–1691)

Most 17th-century scientists who investigated living animals were strongly influenced by the ancient Greek philosophers Aristotle and Galen. William Harvey broke this subservience to the past, and in so doing discovered that blood circulates around the body. This contrasted with the established concept that blood was constantly manufactured in the liver and was used up in the body. Rather than taking the approach of the philosophers, which placed great emphasis on thinking about what might be the case, Harvey formed his opinions after performing experiments and dissections. His interest in anatomy also led him to study the origins of life — the first days of mammalian embryos.

Observing gladiators

Throughout history, blood has fascinated people. Many peoples and religions regarded blood as the seat of life. It made sense, after all: if you drained someone of their blood they had no life left in their body. Some believed that it had almost magical powers, as is shown in the mythical tale of alchemist Dr. Johann Faust, when he seals his bargain with the devil Mephistopheles by signing in blood — the act is seen as signing his life away.

When scientists looked at blood they were faced with two basic questions. Firstly, what did blood do? And secondly, where was it going when it moved in the body's vessels? The second question proved to be the easier to answer; the first took many more years. In the 17th century, most scientists relied heavily on the writings of Aristotle, who believed that things were only present in the body because they had been deliberately put there to perform a particular function. By studying the shape and position of an organ you would be able to establish its function. Its size would give you a good impression of its importance. William Harvey took up this concept and extended it by not only observing the body, but by performing experiments on living animals.

For medical scientists, the other key authority was the Greek physician Galen, who in the second century had spent considerable time working beneath the amphitheatre in Rome. The regular supply of wounded gladiators gave him an unprecedented access to injured people. Galen came up with a complete concept of blood, but sadly one that was completely wrong. He had realized that there were two types of blood-flow from wounds. In one the blood was bright red and came spurting out, and in the other it was dark blue and flowed out in a steady stream. These observations convinced him that these were two different types of blood. He also believed that there was a third form of blood that flowed along nerves.

Galen believed that food was converted into liquid in the stomach. This liquid was passed to the liver where it was turned into blood. This blood flowed out through the body in veins and was used up by the body. Some of the blood went to the heart where it was mixed with air from the lungs and formed a "vital spirit," which then passed into the body through the arteries. Some of this blood ran to the brain and was further refined into "animal spirits." This was distributed through the body by the nerves.

According to Galen, the heart was a furnace that burned oil produced by the liver. The fumes traveled up the pulmonary artery and exited via the windpipe (trachea) — after all, he argued, you could see the fumes on a cold day.

No one seriously questioned this understanding until 16th-century anatomists started reanalyzing the heart. For Galen's concepts to work, there needed to be pores allowing blood to move between the chambers of the heart, but no one could find these pores. Clearly something was wrong.

From flow to circulation

In the mid 16th century, Matteo Realdo Colombo suggested that blood circulates around the body. Colombo also suggested that the heart actively contracts and expels blood, partly via an artery into the lungs and partly via the aorta into the body. The idea was so extreme that few people took it seriously. In any case, Colombo was a surgeon, and proper doctors had little time for these rough-and-

above *Publishing* Anatomica de motu cordis *was a turning point in medical history. It swept away centuries of reliance on Greek philosophy and introduced a new age of science.*

ready technicians. Moreover, Colombo had little hard evidence to support his conclusion. But he did publish the idea in his textbook of anatomy, which Harvey would have seen while studying in Italy.

Having been born and educated in England, Harvey had traveled to Italy to study at the University of Padua. Arriving back in England in the summer of 1602, he settled in London, and by marrying the daughter to the king's physician found himself in the fortunate position of not having to work too hard for a living. This gave plenty of time for research. The increasingly close liaison he formed with King James I of England and then King Charles I, also gave him access to deer from the Royal Estate at Windsor for performing experiments.

One of the problems was that he was either dealing with dead animals, in which the heart was not moving, so it was consequently difficult to work out what it did, or working on living animals, where the heart was beating so fast that he couldn't see its individual movements. He likened the situation to trying to work out how a musket works. The sequence of movements starts with pulling the trigger and ends with the bullet firing out of the barrel. But what happens in between occurs so fast that it is impossible to see.

To get around this, he based many of his observations on the hearts of animals that were dying, the heart consequently beating very slowly, or on snakes that always have slow heart rates. Now he could see that the pulses in arteries came immediately after the heart contracted. He became certain that the pulse was due to blood moving into the vessels.

By careful observation, he found that blood entered the right side of the heart and was forced into the lungs, before returning to the left side of the heart. From there it was pumped via the aorta into the arteries around the body. Crucially, Harvey realized that the amount of blood flowing through this system was too much for the liver to produce. Clearly, the blood had to be circulating back to the veins.

At that time, there were no microscopes, and without these it was impossible for Harvey to see the capillaries that linked the arteries to the veins. Consequently, Harvey was forced to deduce that blood circulated, by carrying out a series of experiments. To start with he took a snake, opened its skin, and squeezed its vena cava — the vein allowing blood to return to the heart. The heart turned white and became smaller, and its beat slowed. It was restored, however, as soon as he took his finger off the vein.

below *William Harvey produced four figures showing the way the valves in the veins control the flow of blood back to the heart. While the hand at upper left has stopped the blood just in front of a valve (at O), the valve prevents the blood from flowing back towards the hand.*

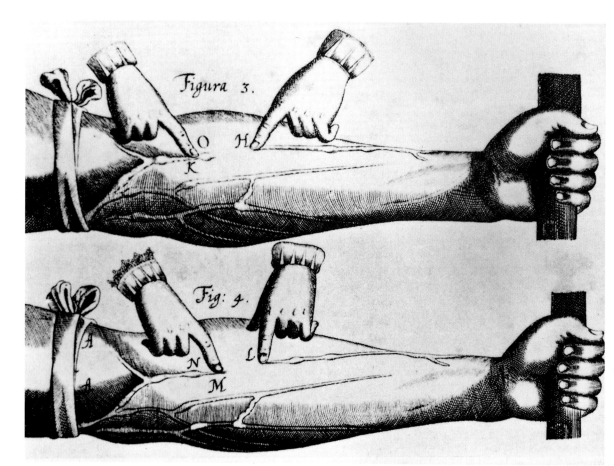

above *William Harvey demonstrating to King Charles I his theory of the circulation of the blood, c. 1616. The English physician was attached to St. Bartholomew's Hospital, and was physician to both James I and Charles I.*

Conversely, if he compressed the artery leading from the heart, Harvey saw the heart swell and turn deep red. He concluded that blood flowed into and out of the heart.

He then showed that blood flowed through arteries in the arm. To do this he tied a bandage tightly around his arm and found that the pulse in the wrist disappeared. He had cut off the flow. Instead, there was a regular throbbing at the site of the bandage, that Harvey attributed to the blood, "Trying to burst through an impediment to its passage and reopen the channel."

He then relaxed the bandage just a little. Blood flowed into the arm but couldn't escape. The veins in the arm filled up and stood out. A 16th-century tutor in Italy, Hieronymus Fabricus of Aquapendente, had recently proposed that the lumps seen in veins were valves, and Harvey correctly deduced that these ensured that blood moved in one direction, flowing back to the heart.

Harvey was convinced that the only explanation of his findings was that blood circulated around the body. He published his findings in 1628. Four years after Harvey's death, Italian biologist Marcello Malpighi discovered capillaries linking arteries and veins. The last piece of the puzzle was in place, though even then no one knew what blood was doing. It would take another 100 years before Antoine Lavoisier discovered oxygen, and worked out what it did in the body.

Seeds of life

Having solved the question of blood, Harvey turned his attention to the beginning of life. As well as identifying valves in veins, Fabricus had made a series of drawings of chick embryos by cracking open eggs on different days of their development. Harvey repeated this work adding new details, and extended it by dissecting pregnant deer from Windsor Great Park.

He concluded that Fabricus had been wrong to assume that a chick is formed from the white of the egg, and decided that it grew from a tiny scab-like blood spot on the edge of the yellow yolk. He then looked for a similar "primordium" in mammals, assuming that they too would grow from some tiny initial speck in an egg. The catch phrase that ran through his second book was "all things from an egg."

Twenty years after Harvey's death, Dutchman Anton van Leeuwenhock used a newly-invented microscope that enabled him to see sperm, and his compatriot Regnier De Graaf realized that mammalian eggs came from ovaries. Harvey would have been impressed.

Timeline

1578
Born on April 1, at Folkestone, Kent
1593–1599
Studies medicine at Caius College, Cambridge
1599
Moves to study at Padua University, Italy
1602
Returns to London
1604
Marries Elizabeth Browne
1607
Elected a fellow of the Royal College of Physicians
1609
Works as a physician at St. Bartholomew's Hospital, London
1618
Takes the appointment of physician to King James I of England
1625
Takes the appointment of physician to King Charles I
1628
Publishes his book *The Motion of the Heart and Blood in Animals*
1629–1633
Travels through Europe with James Stewart, who later became Duke of Richmond
1636
Travels as a diplomat to Germany and Italy
1642
The Civil War forces him to leave London along with King Charles I, and move to Oxford
1645
At the king's request he becomes Warden of Merton College
1647
Returns to London, and after paying fines levied after the Royalist defeat in the Civil War he moves to live with a brother
1651
Publishes his book *The Generation of Animals*
1654
Refuses to accept position as president of the Royal College of Physicians
1657
Dies on June 3 at Roehampton, to the west of London

John
Hunter

1728–1793

Acquaintances
- William Cheselden (1688–1752)
- John Percivall Pott (1714–1788)
- William Hunter (1718–1783)
- Edward Jenner (1749–1823)
- John Bell (1763–1820)

Today's surgeons are highly respected members of the medical profession, but it wasn't always that way. The first surgeons had little anatomical knowledge, but plied their trade because they had sharp instruments and strong arms. They often did surgery in their spare time while working as the local barber or blacksmith. John Hunter has gone down in history as the first scientific surgeon. He started out at a time when people were performing dissections on dead bodies to try and discover how the body functioned and then apply that knowledge to their surgical work.

Strong-arm tactics

The youngest of 10 children, John Hunter had disliked school and showed no interest in learning, but enjoyed looking at insects, birds' nests, and wild animals in general. After his father's death when he was 13 he dropped out of school entirely, and at the age of 17 worked for a few months with his brother-in-law, a timber merchant and cabinetmaker in Glasgow. There he learnt to sharpen tools and use them with care and precision. So when John Hunter moved from Scotland to London at the age of 20 he had all the training needed to become a surgeon—little formal education but skill with a knife, a chisel, and a saw. All he needed now was a little knowledge about the body.

His elder brother, William Hunter, had already established himself as a physician and obstetrician, but at the back of his house also had a dissecting room. John arrived in London just when William was preparing to give his autumn course of lectures in dissection and anatomy. Working in the rooms in Great Windmill Street, John's first task was to dissect an arm. He did it with such precision that his brother soon enrolled him in surgical classes at St. George's and St. Bartholomew's Hospitals. He was also accepted as a pupil of William Cheselden at Chelsea Hospital, and in 1751 started working as an apprentice to John Percivall Pott at St. Bartholomew's Hospital.

William felt that John's lack of learning was an embarrassment, and in the summer of 1755, he persuaded him to enter St. Mary's Hall, Oxford, to learn elocution and classical languages. John found university education just as boring as his childhood school and soon left, eager to get back to the real business of life. He said that he preferred working on dead bodies to learning dead languages.

By the spring of 1761 John Hunter's health was suffering from the poor air in London and the poorer air inside his dissecting room. Looking for a means of getting to a warmer climate, he joined the expeditionary forces and set sail from Portsmouth for Belle-île-en-Mer, a small island off the French coast at the mouth of the Loire. Hunter now had time to employ his dissecting skills in repairing wounded casualties, and used them further as the conflict switched to Portugal. It was a skill that he put to good use for the rest of his life, and recorded his ideas and findings in the book *A Treatise on the Blood, Inflammation, and Gunshot Wounds* that was published a year after his death. Hunter had analyzed many of the problems faced by surgeons and developed rational strategies for dealing with them.

Museums and menageries

At the same time as dissecting human bodies, Hunter started to study other animals, and this became an absorbing passion. While in London he studied all of the different types of fish that he could lay his hands on, and then during his stay in France and Portugal he spent many an hour taking the local animals to pieces.

His interest was not so much to look at animals as members of individual species, but to compare them and look for common structures. Having returned from southern Europe, Hunter bought three pieces of land in Earl's Court and built a house. Soon it was populated with a bizarre collection of animals: some dead, others alive. He had leopards and jackals, buffaloes, stallions, sheep, goats,

above *Hunter broke away from this traditional concept of an anatomy demonstration, where the lecturer sits at a distance and dictates from a book.*

and rams. He planted a mulberry tree to supply leaves for his silk worms and St. John's wort to provide pollen for his bees. Other people enjoyed adding to the menagerie, for example, Joseph Banks (1744–1820) brought him a kangaroo when he returned from James Cook's voyage of 1768–1771.

The rooms of his house were inhabited by mummified bodies, skeletons, cadavers, and dissectors. His collection rapidly became a museum, displaying the way that different species, including human, related to each other. Again, this intricate knowledge of anatomy enabled him to perfect his surgery.

As his fame increased, so did his workload. He liked to start the day in his dissecting rooms with his students, move on to perform surgery, and then finish the day discussing his ideas, or going out to the theatre.

Body snatching

Getting hold of animals was one thing, but finding human bodies to dissect was more difficult. Hunter and his fellow anatomists made use of bodies of people who had been hanged, but even though there were plenty of these victims of rough justice, the anatomists found themselves in need of more.

As any economist knows, where there is a demand, there will be someone ready to supply the appropriate commodity, even if it is human bodies. London had gangs of so-called resurrectionists, who stole bodies from mortuaries, or raised them from new graves. The practice was illegal and met with public outrage. A London publication of 1777 described the conviction of two gravediggers, John Holmes and Peter Williams, who used their job to secure bodies for surgeons. Their punishment was to be whipped as they ran from the end of Kingsgate Street, Holborn, to Diot Street, St. Giles, a distance of half a mile, "and which was inflicted with the severity due to so detestable an offence, through crowds of exulting spectators."

This might have deterred some, but it did not stop the practice. Indeed, in 1783 Hunter is believed to have paid over the odds to get hold of the corpse of an Irish giant who had made his living as a circus attraction. The Irishman's skeleton is still on display in the Hunterian Museum at the Royal College of Surgeons in London.

Timeline

1728
Born in Long Calderwood, East Kilbride, Scotland, on February 13

1748
Moves to London to assist in his brother's dissecting room

1749–54
Studies in London at Chelsea Hospital, St. Bartholomew's Hospital, and St. George's Hospital

1756–63
Serves with the British army in France and Portugal during the Seven Years' War

1763
Works as a surgeon in Golden Square, London

1767
Is elected as a fellow of the Royal Society

1768
Moves to work at St. George's Hospital, London, and starts taking pupils, including Edward Jenner (1749–1823)

1771
Publishes *A Treatise on the Natural History of the Human Teeth*, and marries Anne Home

1773
Starts a series of lectures on "The Principles of Surgery"

1786
Publishes *A Treatise on the Venereal Disease*

1793
Dies on October 16 after suffering heart failure during an argument at St. George's Hospital

1794
Another of his books is published, *A Treatise on the Blood, Inflammation, and Gunshot Wounds*

1749–1823

Acquaintances
— John Hunter (1728-1793)
— James Cook (1728–1779)
— John Hunter (1728–1793)
— Joseph Banks (1744–1820)

Edward Jenner

Infectious diseases have been a scourge of human civilizations, and smallpox was one of the worst. It was highly infectious and killed a third of the people who contracted it. The rest were scarred for life and some were left blind. Physician Edward Jenner decided to fight disease with disease. Following the observation that milkmaids who catch cowpox never suffer from smallpox, he developed a method of infecting people with the cowpox viruses. He didn't understand the underlying scientific mechanism, but it made his patients immune to the disease. The concept of vaccination had arrived.

Fight disease with disease

When Edward Jenner started working as a physician, no one knew about bacteria or viruses, but they were well aware of infectious disease. They knew that there were some diseases that could be passed from person to person. One of these was smallpox — and they don't come worse. When a person became infected they felt at first as if they had influenza, but then a rash developed and spread over their body. A third of adults and up to nine out of 10 children who caught the disease died as the rash developed into pus-filled blisters, followed by infections in the kidneys and lungs.

AN

INQUIRY

INTO

THE CAUSES AND EFFECTS

OF

THE VARIOLÆ VACCINÆ,

A DISEASE

DISCOVERED IN SOME OF THE WESTERN COUNTIES OF ENGLAND,

PARTICULARLY

GLOUCESTERSHIRE,

AND KNOWN BY THE NAME OF

THE COW POX.

BY EDWARD JENNER, M.D. F.R.S. &c.

———— QUID NOBIS CERTIUS IPSIS
SENSIBUS ESSE POTEST, QUO VERA AC FALSA NOTEMUS.
LUCRETIUS.

London:

PRINTED, FOR THE AUTHOR,

BY SAMPSON LOW, Nº. 7, BERWICK STREET, SOHO:

AND SOLD BY LAW, AVE-MARIA LANE; AND MURRAY AND HIGHLEY, FLEET STREET.

1798.

above Getting new ideas accepted is never easy, so it is not surprising that Jenner had to self-publish his ideas about vaccination in order to get them noticed. His discovery that cowpox innoculation gave an immunity to smallpox introduced the idea of vaccination against disease into medicine.

Writing in the 1840s, English historian Thomas Macaulay painted a graphic picture of the disease: "The smallpox was always present, filling the churchyards with corpses, tormenting with constant fears all whom it had stricken, leaving on those whose lives it spared the hideous traces of its power, turning the babe into a changeling at which the mother shuddered, and making the eyes and cheeks of the big hearted maiden objects of horror to the lover."

There was only one way of becoming immune to the disease, and that was to have it and survive. This "variolation" wasn't a pleasant option, but some people deliberately exposed themselves, or their children, to small doses of the disease in the hope of getting a mild infection and living to tell the tale.

In China, a practice developed of giving a controlled dose of the illness by grinding smallpox scabs and blowing a small amount of the dust into people's nostrils. Then, in the 1600s, the Chinese developed a method that involved swallowing pills made from fleas that had been living on cows contaminated with cowpox. These pioneers didn't know it, but they had just invented oral vaccination.

From variolation to vaccination

Variolation arrived in Europe in the early 1700s after an English aristocrat, Lady Mary Wortley Montague, came back from a trip to Turkey. She had a personal interest in smallpox as it had badly scarred her when she was young and had killed her brother at the age of 20.

On her visit she had been intrigued to see women in the Ottoman court making small graze marks on children's arms and wiping the area with smallpox scabs. The children subsequently became immune to the disease. Convinced that this was a life-saver, she had her five-year-old son inoculated in Turkey, and got a surgeon to perform the routine on her four-year-old daughter as soon as she returned to England in April 1721. This was carried out before the royal court physicians, who then repeated the experiment on six prisoners who were granted a Royal pardon if they survived. They did, and were duly released.

While the technique was fairly successful, it sometimes triggered full outbreaks of the disease and one in 50 people would die as a result of the variolation. There had to be a better solution.

Then, in 1796, English physician Edward Jenner performed a vital experiment. Jenner was the third child of Reverend Stephen Jenner, vicar of Berkeley, Gloucestershire. At the age of 13 he had become an apprentice to eminent surgeon Daniel Ludlow (based in Sodbury, near Bristol), and joined surgeon John Hunter when he was aged 21. Two years later he returned to work in Gloucestershire, and a further 24 years passed before he made his mark on history.

On 14 May 1796 he made two half-inch incisions on the surface of eight-year-old James Phipps' arm and wiped a cowpox scab over the wound. The scab had originally come from the hands of Sarah Nelmes, a local milkmaid. Folklore claimed that milkmaids who caught cowpox didn't get smallpox. Six weeks after giving the boy cowpox, Jenner deliberately exposed him to smallpox. He did not become infected. Without knowing it, he had repeated the Chinese discovery.

The process had worked for the Chinese and for Jenner because, unknown to them, the virus that causes cowpox is remarkably similar to the one that causes smallpox. But, while cowpox will not cause a severe disease in humans, it will enable their bodies to develop resistance to both cowpox and smallpox.

The idea, however, was controversial and met severe opposition. The Royal Society rejected the manuscript that he wrote describing the work, on the basis that it was "in variance with established knowledge" and "incredible." Jenner was not to "promulgate such a wild idea if he valued his reputation." In the end he published the paper himself, and a year later over 70 principal physicians and surgeons in London signed a declaration expressing their confidence in the idea.

Vaccination was born, and intriguingly our language still remembers the event — the word "vaccinate" has its roots in the Latin *vacca*, meaning "cow."

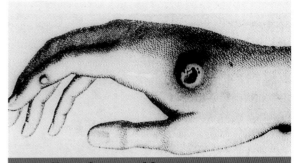

A smallpox-free world

By 1800 some 100,000 people throughout the world had been vaccinated against smallpox. In North America, President Thomas Jefferson was one of the first key proponents of the method, appointing Benjamin Waterhouse, professor of the "theory and practice of physic" at Harvard Medical School, as Vaccine Agent in the National Vaccine Institute. In 1805 Napoleon vaccinated all his troops who had not had smallpox and insisted that all civilians were vaccinated a year later.

In more recent history the World Health Organization mounted a worldwide mass-vaccination program. This weakened the disease's hold within whole populations, and medical rapid-reaction forces pounced on any minor outbreaks. On March 2, 1971, two children were brought into the Eduardo Rabelo Hospital in Rio de Janeiro. They were the last recorded cases of smallpox in the Americas. The last known case of naturally caught smallpox occurred on October 27, 1977, in a small Somali village called Merka, when Ali Maow Maalin, a 23-year-old hospital cook, took an infected child for treatment. Maalin caught the disease, but survived. The child died.

Since then, the only person to die of smallpox has been a laboratory photographer, who worked next to a place where smallpox viruses were stored. Thanks to Jenner and vaccination, smallpox now only exists in laboratories — in the wild it has been consigned to history.

below *The smallpox virus has a protein caat (green) surrounding its DNA (red) genetic material. Smallpox was eradicated in the 1970s but isolated cultures of the virus are still kept in laboratories for research purposes.*

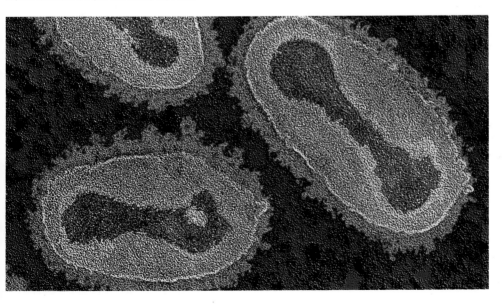

People with injury or disease are often weak and fighting for their lives. They are therefore particularly vulnerable and likely to catch other diseases. Placing them in a hospital may bring them to a place where there is medical expertise, but unless the hospital is scrupulously clean it could put the patient at increased danger. Florence Nightingale was the first person to collect data about diseases in hospitals and realize that the basic need of a hospital was good sanitation. In doing this she revolutionized healthcare and turned nursing into an esteemed profession.

Florence Nightingale

1820–1910

Acquaintances
— Sydney Herbert (1810–1861)
— John Delane (1817–1877)
— Elizabeth Blackwell (1821–1910)
— Josephine Elizabeth Butler (1828–1906)
— Elizabeth Garrett Anderson (1836–1917)
— Sophia Louisa Jex-Blake (1840–1912)

War and welfare

Illness is never a desirable idea, and being in hospital should be seen as a last resort. While hospitals are places where there is a concentration of medical knowledge, they are also places where there are concentrations of sick people. Consequently, one of the best places to catch a disease is in a hospital. And that is the case for a modern hospital, even though the need for cleanliness and sanitation is thoroughly understood.

When Florence Nightingale entered nursing in the 1850s, the causes of disease had not been identified, and hospitals were desperate, stinking places filled with people who had missing limbs and running sores. There had been little attempt to create systems of management and practice that could help prevent disease passing from patient to patient. Indeed, the fact that diseases could also readily pass from patient to carer was one of the reasons why nursing was seen as an occupation for women from the lower classes.

At the age of 17, Florence Nightingale became convinced that she had a calling from God to some as yet unidentified great cause. Then, aged 25, and despite her parents displeasure, she decided to head for nursing. Her father was a passionate campaigner for people's rights, and had taken care to see that Florence had a good education. She was fluent in Italian, Latin, and Greek, and was well versed in history and mathematics. He had hoped that she would move on to higher things than nursing. Nightingale, however, used her education to revolutionize nursing.

Her big break came during the Crimean War. Appalled by the reports of conditions in the makeshift hospitals in what is now Turkey, she wrote to the War Office asking to be allowed to go and sort the situation out. Arriving at Scutari with 38 nurses, she found that the men were being kept in rooms without blankets or decent food, and that they were still dressed in the uniforms in which they had been wounded. These, she reported, were "stiff with dirt and gore." Diseases such as typhus, cholera, and dysentery were rife, and these diseases alone were responsible for one in six of the deaths that occurred during the war.

Shame and sanitation

Realizing the power of public opinion at home, Nightingale sent reports of the festering conditions to a contact at *The Times* newspaper in London. Shamed into action, the army started to clean up the hospitals and make the soldiers more comfortable, but the deaths continued. Sadly, the hospital was built above a system of blocked and defective drains, and even Nightingale didn't realize that disease could rise in the air from the sewers and infect her patients.

She did, however, collect data, and when she returned to London in 1856 she sat down and started to analyze it. This was the first time that anyone had taken a systematic look at patterns of disease within communities.

One of her first realizations was that there was a clear pattern linking the numbers of people dying of diseases and the seasons. By the time she presented her data to the 1857 Sanitary Commission, she had developed a polar area diagram that graphically displayed these patterns. Through

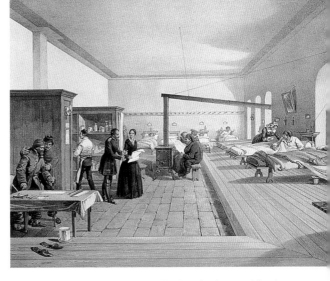

above *Initially Florence Nightingale thought that good food was the essential ingredient to nursing. She later decided that it was ordered hospitals and basic hygiene.*

studying her data she changed her opinion that good food was the key issue in recovery, and became convinced that the most basic need in a hospital was good sanitation.

She had combined a concern for people with a careful application of statistics, and created the foundations of the modern concepts of nursing and medical care.

Timeline

1820
Born in Florence, Italy, on May 12
1849
Travels to Europe to study different ways of running hospitals
1850
Begins a training in nursing at the Institute of St. Vincent de Paul in Alexandria, Egypt
1853
Becomes the superintendent of the Hospital for Invalid Gentlewomen in London
1854
Works in Turkey during the Crimean War
1858
Becomes a fellow of the Royal Society of Statistics
1859
Publishes *Notes on Hospitals*
1860
Nightingale uses funds raised in tribute to her services to found the Nightingale School and Home for Nurses in London
1907
Becomes the first woman to receive the British Order of Merit
1910
Dies in London on August 13. Five years later, the Crimean Monument in Waterloo Place, London, is erected in her honor

Louis Pasteur

1822–1895

Acquaintances
—Antoine Jérôme Balard (1802–1876)
—Theodor Schwann (1810–1882)
—Joseph Lister (1827–1912)
—Louis Duclois du Hauron (1837–1920)
—Robert Koch (1843–1910)

Throughout history, people had generally accepted that some life forms came into being spontaneously from non-living matter. Maggots, for example, appeared on decaying meat without any apparent cause. Similarly, there was the opinion that diseases could come from nowhere and attack people. Louis Pasteur's revolutionary discoveries started with the realization that living micro-organisms produce crystals that have different properties from those generated by purely physical processes. This led him to realize the importance of microbes in brewing, and that microbes were the cause of many diseases. A chance event led him to discover that he could also produce vaccines to combat these diseases.

Crystals give clues

Like so many geniuses, Louis Pasteur showed little sign of intellectual superiority at school. As the only son of a poorly educated tanner, he preferred fishing and drawing to schoolwork. His father managed to keep him in the local school, and toward the end of his time there Pasteur became intrigued by chemistry. His head teacher was sufficiently impressed with the way that Pasteur approached the subject to encourage the teenage boy to apply for a place at the prestigious Ecole Normale Supérieure in Paris. He applied, was accepted, and was set on his path of discovery.

While Pasteur is best known for his work in the realms of microbiology, he started his scientific life as a chemist. Working in the laboratory of Antoine Jérôme Balard, he studied the emerging science of crystallography and started looking at tartaric acid. The crystals of this organic acid often form in the sediments of fermenting wine, along with crystals of a remarkably similar compound known as racemic acid. Scientists were confused by these two compounds, because recent data had shown that they were chemically identical, yet they had strikingly different properties if you dissolved the crystals in water and shone polarized light through the solutions. In the case of tartaric acid, the light beam was rotated to the right, while nothing occurred to the racemic acid. Clearly the two were not identical.

Pasteur examined the crystals under his microscope, and suddenly spotted that the racemic acid crystals occurred in two different shapes. The two were mirror-images of each other. When he painstakingly sorted the crystals into separate piles, he found that one pile rotated light to the left, the other to the right. Unwittingly, Pasteur had invented the science of stereochemistry.

It was the next step of logic that was the key to the rest of Pasteur's life: he decided that these two different types of crystals were present because something living had made them. They were not simply the products of an isolated chemical reaction.

Then, in 1854, Pasteur was asked to help a company that produced alcohol by fermenting beet root. The company had a problem. Rather than getting alcohol they often produced lactic acid. At this time, chemists were convinced that sugar turned into alcohol, carbon dioxide, and water because of some inherent "unstabilizing vibrations." You could make a fresh vat of munched-up roots ferment by transferring a small quantity of these unstabilizing

vibrations from a finished vat. They knew that yeast cells were present in the vats, but thought that they played no role. Any scientists who suggested that the yeast was important were ridiculed by the established elite, who were certain that living systems had been driven out of the pure discipline of chemistry.

Arriving at the factory with his microscope, Pasteur found that all the vats producing alcohol were inhabited by healthy plump yeast cells. Those vats producing lactic acid had small rod-like microbes mixed in with less-robust yeast. He then discovered that alcohol was not the only product in the vats, and there were traces of other compounds that could not have come from a simple breakdown of sugar. When he found crystals of compounds, some of which rotated light to the left and others to the

above *This flask was used by Louis Pasteur in his experiments which proved that tiny organisms in the air can cause decay. His work led to the introduction of antiseptics. The flask has remained sterile since it was heated and sealed in the 1860s.*

right, he was convinced. Fermentation was a living process—yeast made alcohol, and the rod-like organisms that we now call bacteria were souring the fermentation.

Over the next few years Pasteur discovered that if he heated the liquor from the vats he could kill the living micro-organisms and stop all reactions. He had "pasteurized" the batch. This discovery eventually revolutionized the safety of foods, in particular milk, where pasteurization kills bacteria and has radically reduced the spread of food-borne disease.

The death of spontaneous generation

In the 1860s, at the time that Pasteur was analyzing fermentation, there was another debate raging. Could beetles, maggots, eels, mice, and other forms of life spontaneously appear? The idea had started with Aristotle, and most people firmly believed in it. For example, a 17th-century recipe for the spontaneous production of mice recommended that you place sweaty underwear and husks of wheat in an open-mouthed jar and wait for around 21 days. The sweat, it was claimed, would penetrate the husks of wheat and change them into mice.

His experience with fermentation led Pasteur to believe that nothing appeared spontaneously. He was also aware of experiments that had been performed almost 200 years earlier by Italian naturalist Francesco Redi. Redi had taken two lumps of meat and placed them in separate jars. One jar he left open to the air, and the other he covered with gauze. Maggots appeared only on the meat in the open jar, causing Redi to conclude that they had not arisen spontaneously, but had come because adult flies had laid eggs on the meat.

Pasteur performed a similar set of experiments, but this time with grape juice. He drew the juice out of a series of grapes using a heated hypodermic needle, and placed it in a sealed jar. The juice didn't ferment. But juice that had some dust from the grapes' surface did ferment. The yeast, concluded Pasteur, arrived on the dust—it didn't appear spontaneously.

In his clinching experiment, he placed juice in a round flask with a long thin neck. He heated the neck so that he could bend it down and up in an "s" shape, before sealing the end. Nothing grew in the flask. Next he snapped off the seal. Again nothing grew. Dust did fall into the open end, but it couldn't reach the liquid. He tilted the flask so that the liquid washed over the dust and then returned to the

above *Pasteur's name is remembered by everyone who consumes processed dairy products as the milk is almost always carefully heated and cooled as it is pasteurized.*

flask—the liquid started to ferment. Pasteur had shown that it was not access to the air, but the presence of dust that triggered the process. "Never will the doctrine of spontaneous generation recover from the mortal blow of this simple experiment," said Pasteur. "No, there is now no circumstance known in which it can be affirmed that microscopic beings come into the world without germs, without parents similar to themselves."

Germ theory and vaccination

In another episode of his life, the French Department of Agriculture asked Pasteur to investigate a disease that was devastating silkworms. He taught the silk farmers to recognize silkworms that were free of the two key diseases, pebrine and flacherie, and helped them set up breeding programs using only disease-free animals.

During his research, he found that healthy worms could catch the disease if they fed on leaves that had previously been in contact with sick worms, and that control of the environment and the quality of food could restrict the disease's ability to spread through a population.

Added together, Pasteur's work was leading him to realize that diseases were caused by micro-organisms—by germs. In England, the pioneering surgeon Joseph Lister was beginning to incorporate the ideas into his techniques and insisted on using clean instruments and bandages as well as spraying phenol solutions into the air around an incision.

One opponent of health was cholera, and Pasteur spent many years studying the form that affects chickens. He had identified the bacterium and was growing it in culture. If he injected it into a chicken, the bird died within 48 hours.

Then he went away from the lab for a few weeks one summer and returned to find that his cultures of cholera could no longer kill chickens. They didn't even make the birds sick. Initially Pasteur and his colleagues were disappointed—maybe these bacteria weren't the culprits after all? They grew a fresh batch of bugs and injected them into the chickens again. Those that had received the old bugs a few weeks earlier were unaffected. Those that had received only the new batch died.

Pasteur realized that he had come across this sort of thing before. It was the same as Edward Jenner's discovery about smallpox. Living in the culture for so long must have weakened the bacteria. Far from causing disease, the bugs now enabled chickens to develop immunity. He had a vaccine to the disease.

Soon Pasteur realized that if he could do it with cholera, then he could surely do it with other diseases. He had not only discovered the origins of disease, but also revealed a potent weapon. When he died on September 28, 1895, the world was a safer place.

top right *Two dividing yeast cells of the fungus* Candida albicans *which causes human thrush. Here, this spore (yeast) stage is a non-infective reproductive stage. From spores, the infectious "mycelial" form develops. Tubular filaments branch into a spreading network which then produces more spores.*

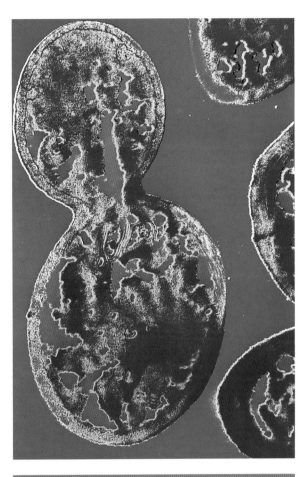

Timeline

1822
Born in Dole, France, on December 27
1847
Receives his Ph.D. in chemistry
1854
Takes appointment as dean and professor of chemistry at the Faculty of Sciences in Lille, France.
1856
Starts investigating problems in the alcohol-fermenting industry
1857
Becomes the director of scientific studies at the Ecole Normale Supérieure, Paris
1859
Produces evidence that destroys the theory of spontaneous generation of life
1865
Turns his attention to tackling diseases in silkworms
1867
Becomes professor of chemistry at the Sorbonne, Paris
1888
Becomes the first director of the Pasteur Institute
1895
Dies on September 28

When Koch began his career, scientists were beginning to understand the association between bacteria and various diseases. Koch developed methods of staining bacteria that enabled him not only to see them under a microscope, but also to differentiate between the various strains of micro-organism that he found. Koch now proved that specific organisms caused specific diseases and, in addition, pollution could spread disease. He developed methods for obtaining pure cultures of bacteria and laid down four conditions, Koch's Postulates, that need to be satisfied to be sure that a particular type of bacteria causes disease.

Robert Koch

1843–1910

Acquaintances
— Louis Pasteur (1822–1895)
— Richard Julius Petri (1852–1921)
— Paul Ehrlich (1854–1915)
— Emil von Behring (1854–1917)

Pure bacteria

In the mid 1800s scientists were coming to terms with the concept that something as small as a bacterial cell could kill an animal as large as a human or a cow. Then, in 1876, Robert Koch announced that he had isolated the species of bacteria responsible for anthrax.

Koch had trained as a doctor in Germany, and after his wife bought him a microscope for his 28th birthday he started looking at microbes as a pastime. He had taken drops of blood from animals infected with anthrax and sandwiched them between two microscope slides. Through the lens he could see the numbers of bacteria grow as the cells constantly divided. Staring down the microscope, he saw that as the blood dried out the cells changed. They formed small white capsules. Koch discovered that these "spores" protected the bacteria and could survive in a dormant state for years. If injected into an animal, the spores would wake up and trigger a fresh growth of bacteria—a new infection.

This combination of findings explained some curious aspects of the disease. The farmers had been frustrated that when they turned cattle out onto pastures in the spring many animals became sick and died. If the cause of the disease was really bacteria, how had the tiny organisms survived in the fields through the icy winter? Now they had the answer. The bacteria survived as spores.

Some scientists were unconvinced. They thought that something in the initial sample other than the bacteria must have been triggering the disease. This issue was resolved when French bacteriologist Louis Pasteur extended Koch's work by taking a small sample of bacteria from a flask of anthrax and placing it in a new flask. He allowed the bacteria to multiply before repeating the procedure. After 100 successive transfers to clean brewing solutions, Pasteur was convinced that the only thing present must be the anthrax bacteria. This culture was still capable of causing anthrax. The bacteria must be the killer.

A chance observation caused the next step forward. Koch noticed that a slice of old potato left on a bench was covered in spots of different colors, and realized that each spot was a colony of a different type of micro-organism. Up to this point, scientists had grown bacteria in flasks, or injected them into animals. Koch realized that growing them on solid surfaces would mean that they could easily see any contamination. His colleague Richard Julius Petri designed a shallow flat dish and filled it with a thin layer of culture medium that he set solid by adding gelatin to the mix, and so the Petri dish entered service in microbiology laboratories.

Koch's Postulates

The more Koch studied bacteria, the more he realized that some were friendly, but others caused disease. He was keen to establish a set of rules that would help identify the killers.

His list of criteria became known as Koch's Postulates:
- The micro-organism must be identified and seen in all animals that suffer the same disease.
- It must be cultured through several generations.
- These later generations of bacteria must be capable of causing the disease.
- The same agent must be found in the newly infected animal as was found in the original victim.

Using this set of criteria he identified the specific bacilli that caused some 20 or more diseases, including tuberculosis, cholera, salmonella, pneumonia, and meningitis.

Koch's desire was not just to identify the killers, but to find ways of combating them. He spent a huge amount of time working on a vaccine for tuberculosis, and had very little success. This failure needs to be set in the context that almost a hundred years after his death, scientists have still failed to find a vaccine that enhances the body's ability to fight this particular type of bacteria in adults, although the current "BCG" vaccine has reasonable effect in children. Twenty-first century science may have a clearer view of the enemy, but finding the right weapon has proved difficult.

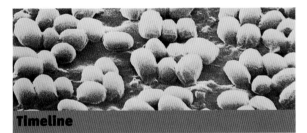

Timeline

1843
Born in Hanover, Germany, on December 11
1866
Receives his M.D. degree from the University of Gottingen
1867
After training at Hamburg General Hospital, he starts general practice at Langenhagen
1870
Volunteers for service in the Franco-Prussian War
1876
Discovers the type of bacterium that causes anthrax
1882
Discovers the type of bacterium that causes tuberculosis
1883
Travels to Egypt as leader of the German Cholera Commission and discovers the bacterium that causes cholera
1885
Becomes director of the Institute of Hygiene in Berlin
1891
Becomes the director of Berlin's Institute for Infectious Disorders
1893
Formulates rules for controlling epidemics of cholera and other diseases
1905
Receives the Nobel Prize for Physiology or Medicine
1910
Dies at Baden-Baden, in Germany, on May 27

In the late 1800s the words cholera, typhoid, tuberculosis, and diphtheria cast fear throughout the world. It was a reaction that Paul Ehrlich hoped to change as he sought for and found scientific ways of combating the diseases. Ehrlich was convinced that he could find "magic bullet" chemicals that would specifically kill bacteria without harming the person infected by the bacteria. He found chemicals that could treat sleeping sickness and syphilis, and worked out how the body's natural defense system combats disease-causing bacteria. He referred to the concept of fighting diseases with chemicals as "chemotherapy."

Paul Ehrlich

1854–1915

Acquaintances
— Robert Koch (1843–1910)
— Ilya Ilyich Mechnikov (1845–1916)
— Emil Behring (1854–1917)

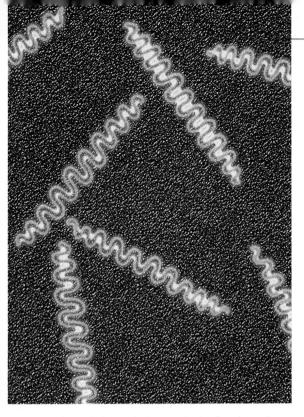

above *These spirally-twisted, elongated spirochete bacteria cause syphilis in humans.*

Inspired by Koch

In 1882, Paul Ehrlich attended a lecture given by the microbiologist Robert Koch. Koch was talking about the bacillus that causes tuberculosis, stating how difficult it was to recognize. Inspired by what Koch had to say, Ehrlich set out to solve the problem. What was required was a stain that bound to these strange bacteria. A few months later he had tried hundreds of different dyes and found one that stuck to the bugs and marked them out. It also marked him out as a problem solver.

A few years later he started working with Koch at the Institute for Infectious Diseases in Berlin, and along with Emil Behring turned his attention to diphtheria. Researchers in Paris had discovered that the disease was not caused by the diphtheria bacteria, but by chemicals released from the bacteria. These toxins destroyed blood cells, and consequently killed the infected animal. But close observation showed that not all animals died. Some animals were immune, and Behring realized that they were producing a chemical that could neutralize the bacterial toxin—they were making an anti-toxin. Furthermore, he found ways of isolating the anti-toxin and using it to treat people affected by diphtheria.

This got Ehrlich thinking about the way that toxins work. He came up with the idea that part of the toxin molecule must lock onto cell walls in the host, while another part of the molecule causes the damage. In addition, he suggested that if the cell survived the attack it would produce more receptors and let them fall off into the blood stream. The toxin would then bind to these free-floating receptors, or antibodies, and consequently not be able to bind to any cell.

Dyes and magic bullets

In 1896, Ehrlich turned his attention to searching for "magic bullets." His theory was that if the body could produce antibodies that neutralized toxins, there must be a way of making chemicals artificially that would have a similar effect. He came across the work of a British army scientist who had identified the micro-organisms that caused sleeping sickness—they were a group of single-celled protozoa called the trypanosomes, which live in their victim's blood.

Ehrlich and his Japanese assistant, Kiyoshi Shiga, found that one particular red dye was capable of destroying these parasites in the laboratory, though sadly its effect in medical use was less dramatic. It did, however, prove that the concept of finding killer chemicals was viable. His next target was the bug that causes syphilis, a spirochete. It had recently been discovered by two other researchers, Schaudinn and Hoffmann, and Ehrlich was determined to find a chemical that would kill it. At the time he was working to develop a series of compounds that incorporated arsenic, and he started testing them one by one. The first 605 didn't have any effect, but compound 606 did.

After performing hundreds of experiments to make sure that the chemical really did kill the syphilis bug but didn't harm humans, he announced the new wonder drug, calling it Salvarsan. The drug worked, but was difficult to manufacture, so Ehrlich went back to the laboratory and started looking for another. Compound 914 proved to be the winner. It was not as potent but was much easier to make, and so Neosalvarsan was launched on the market. His search for "magic bullets" had led to an era of "chemotherapy"—chemicals that were designed to cure.

Timeline

1854
Born in Strehlen, Upper Silesia, Germany, on March 14

1878
Receives his doctorate of medicine

1882
Publishes his methods for staining the bacteria that Robert Koch had identified as being the cause of tuberculosis

1890
Koch invites Ehrlich to move to the newly established Institute for Infectious Diseases in Berlin

1896
Becomes the first director of the Institute for the Control of Therapeutic Sera, in Berlin

1908
Shares the Nobel Prize for Physiology or Medicine with Ilya Ilyich Mechnikov (1845–1916)

1914
Has a stroke caused partly by the stress he suffered at the outbreak of World War I

1915
Dies of a second stroke while on holiday in Bad Homburg, on August 20

Alexander Fleming

1881–1955

Acquaintances
—Howard Walter Florey (1898–1968)
—Andrew J. Moyer (1899–1959)
—Ernst Boris Chain (1906–1976)
—Norman Heatley (1911–)
—Charles Fletcher (1911–1996)

Once scientists had found that bacteria caused disease, the race was on to find ways of killing bacteria without harming the patient. While research in Germany focused on developing dyes that could stick to the bacteria and damage them, a chance discovery in England opened a new avenue of research. Alexander Fleming noted that a chemical produced by mold could kill bacteria. He was the first to identify an antibiotic, but it took others, including Ernst Boris Chain and Howard Walter Florey, to develop its use in medicine. For the first time in medical history, human beings had a way of fighting bacterial infections.

The first patient died, but antibiotics lived on

On December 27, 1940, 48-year-old London policeman Albert Alexander lay in Briscoe Ward in Oxford's Radcliffe Infirmary. He had scratched his face earlier in the week either on the thorn of a rose or while shaving. But whatever the cause, the scratch had become infected. Bacteria and toxins were flooding into his blood and he was close to death. The situation was shockingly common, and he lay in a ward filled with similarly afflicted patients. While drugs containing various forms of sulfur had some affect on bacterial infections, they didn't help Alexander. On February 3, doctors removed one of his eyes and drained abscesses in an effort to combat the disease. The other eye and his lungs were also infected.

Then, on February 12, hospital clinician Dr. Charles Fletcher arrived in the ward with 200 milligrams of crudely purified penicillin and injected doses once every three hours. By the next day Alexander's temperature was normal and he was able to sit up and eat. Sadly, the relief was short-lived. A few days later the stock of penicillin ran out, the disease re-established itself and on March 15 he died. Even though the first patient didn't live to tell the tale, scientists realized that they had discovered a powerful new weapon.

A natural defender

Unlike the magic bullets that German chemist and microbiologist Paul Ehrlich hoped to create, antibiotics have always existed in the biological world. They are chemicals produced by one micro-organism in order to prevent another organism competing against it. They enable an organism to create space for itself in an overcrowded environment.

This was what Alexander Fleming famously spotted on a culture dish in September 1928. He had been called back from a holiday in Suffolk to come into St. Mary's Hospital, London, and assist a colleague who was treating a patient with a bad abscess. Fleming's attention was drawn to an unusual pattern of growth in one of the Petri dishes sitting on a bench in his laboratory. At one edge was a mold, and on the other side of the dish were colonies of *Streptococcus* bacteria. But in a circular zone around the mold the bacteria had died, they had broken apart— or "lysed" as bacteriologists call the process.

Fleming realized that he had previously seen bacteria lyse a few years earlier when a tear had fallen onto a Petri dish. He had also seen that some bacteria were unable to grow next to a fungus. But he had never seen disease-causing bacteria lyse because they were near a fungal colony. This wasn't just preventing bacteria from growing, it was killing any that were present.

The mold was unusual and had probably entered Fleming's laboratory via a connecting shaft of a dumb waiter that linked his room to that of C.J. LaTouche. This Irish scientist was attempting to show that fungi caused asthma. LaTouche incorrectly told Fleming that the mold was *Penicillium rubrum*—a type of mold that we now know does not produce penicillin.

Fleming realized that if his discovery was to be useful, he needed to find a way of purifying the compound, and finding himself unable to do this he devoted his attention elsewhere. Thankfully, he had lodged a sample of the mold in the collection held by the Medical Research Council, so when other people began working on it, they used the correct mold, but to start with just gave it the wrong name. The correct identity turned out to be *Penicillium notatum*.

above *While working at St. Mary's Hospital in London, Fleming recorded the famous plate on which he noted that mold was actively destroying colonies of bacteria. The rest is history...*

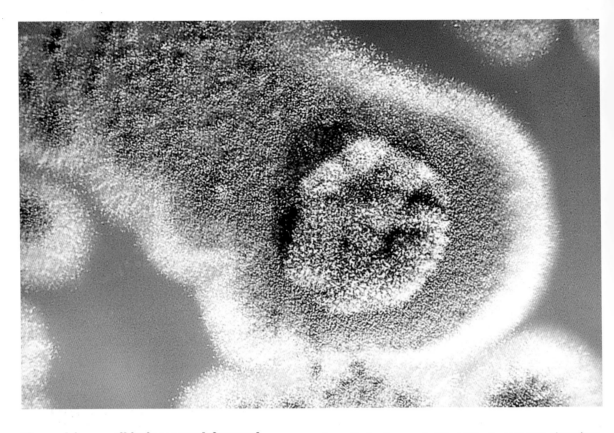

Florey drives antibiotic research forward

In 1922, 18 years before the first use of antibiotics, 23-year-old Australian-born Howard Walter Florey left his home in Adelaide to take up a Rhodes Scholarship in Oxford. Colleagues described him as uncompromising, energetic, and rather prickly — qualities that proved useful over the next few years. After excelling at Oxford, he moved to Cambridge and from there to Sheffield University. Then, aged 37, he became the professor of pathology at the Sir William Dunn School in Oxford.

With few funds, Florey built up an interdisciplinary team comprising chemists, pathologists, bacteriologists, and physiologists. It was a radical concept at the time. Team member Ernst Boris Chain was a chemist, and set about extracting penicillin. Having succeeded, he injected it into a rabbit to see what would happen. If animals weren't harmed, but bacteria died, maybe this bug-killer could cure infections? The rabbit seemed unaffected.

He now needed larger quantities of the compound so that he could see what happened to bacteria in animals. To create this, the Oxford team had set up a Heath Robinson penicillin factory, and used bedpans as fermentation flasks. On Saturday May 25, 1940, biochemist Norman Heatley, Chain, and Florey performed a ground-breaking experiment. They injected eight mice with 110 million streptococci, a virulent strain that should kill within a day. One hour later, they gave penicillin to four of the mice. The other four received nothing and acted as controls.

Heatley watched and waited. By late afternoon the four control mice were sick and by 3.30 a.m. all were dead. In

above **Photomicrograph of** Penicillin chrysogenum, *three days old, grown on potato dextrose agar medium (yellow).*

contrast, the four treated with penicillin were healthy. Heatley cycled back to his rooms through wartime, blacked-out Oxford to catch a few hours' sleep before telling his supervisor, Florey.

The experiment had been a success, but they were again faced with a problem of scale. Drugs are almost always given in a dose that relates to the body mass of the recipient, because the drug spreads through the body. The bigger the animal, the more drug would be needed. A mouse is 3,000 times smaller than a human being, but their mouse experiment had used up all their stocks of penicillin. To be of any use, they were going to have to expand the operation on a massive scale.

By now the Battle of Britain was in full swing, and with the German Luftwaffe bombing English cities the team struggled to set up a new production system. Eventually Florey managed to convert a section of the Dunn School into a factory and filled it with ceramic fermentation vessels.

In the mean time Heatley devised an extraction and purification system for penicillin. He acidified the solution containing the drug and then poured in some ether. When the two liquids were shaken, the penicillin left the acidic solution and went into the ether. Most of the impurities, however, stayed in the watery acid. Stop shaking, and the two liquids separate. But now Heatley needed to get the penicillin out of the ether. This he did by shaking the ether with water that was held at a neutral pH. The penicillin came back into the clean water.

As February 1941 arrived they believed that they had enough penicillin to try treating a human and chose to start with Albert Alexander. With his death the team realized they would need vastly more.

America joins in the race

As often happens, the rest of the story of penicillin's routine use in medicine is complicated by international rivalry, personality clashes, and good luck.

In July 1941, Florey and Heatley flew via neutral Portugal to New York and started collaborating with people at the Department of Agriculture's North Region Research Laboratories at Peoria, Illinois. Heatley set to work with Andrew J. Moyer. Initially they worked together, but Heatley found that Moyer became increasingly secretive and when Heatley returned to England in July 1942 he discovered Moyer had published their work, omitting to include Heatley's name on the publication. When, subsequently, Moyer applied for a patent he used the publication to claim his right as the sole beneficiary. This simple move meant that all the profits from the discovery of penicillin went to America, with the original pioneers receiving only international recognition for their work.

However, before that could happen, they needed to increase the scale of production. Heatley and Moyer worked on a project growing mold on the liquor that was left after starch was extracted from corn. The overall hope was to get something useful out of this waste product. Fleming's mold grew on this liquor, but the quantities of penicillin produced were still small.

The December 1941 attack on Pearl Harbor caused America to redouble its efforts for the war. The authorities saw that penicillin could save the lives of wounded soldiers, and military personnel were ordered to gather handfuls of soil from around the world in the hope of tracking down a fungus that produced high quantities of penicillin. In the end, the army was beaten by Mary Hunt, a laboratory tea lady, who had become known as "moldy Mary" because of her enthusiasm for collecting bits and pieces from rubbish bins. One day she arrived with a lump of rock melon infected with *P. chrysogenum*, a strain that produced 3,000 times more penicillin than Fleming's original mold.

Soon American factories were making billions of units of penicillin a month and the money generated has underpinned the U.S. pharmaceutical industry pretty well ever since. By 1979 some 15,000 tons of fermented bulk product were produced, having a minimum market value of 240 million dollars, that when purified were sold for 1,250 million dollars.

War and recognition

The arrival of antibiotics toward the end of World War II became a major factor in the war. During World War I, almost one in five of the American army was killed by pneumonia. This fell to less than 1 percent in World War II. Military commanders were also impressed that one or two

injections of penicillin cured gonorrhea. As the war progressed they had found themselves with thousands of troops who were incapable of fighting because of the disease. The army was also in desperate need of troops to invade Sicily. Florey, however, was not keen to see his precious drug used to salvage people who had acquired sexually transmitted diseases. General Poole asked Winston Churchill what to do and received a reply that "this valuable drug must on no account be wasted. It must be used to the best military advantage." Poole considered that treating these men was of great military concern. They got penicillin and were returned to their units.

Despite the fact that Fleming played little part in making penicillin available, his name has gone down in history as the key player in the story. Part of the reason is that in August 1942 a friend of Fleming's was dying of bacterial meningitis. Fleming had tried treating him with sulfur drugs, but had no success. He then phoned Florey asking if there was any chance of using some of his penicillin. Florey handed over all the available stock and the friend's life was saved.

Fleming was so excited by this that he announced it to the press. Wearing a white coat, he had his photo taken and he sent the press to Oxford to get the rest of the story from Florey. Knowing how little penicillin they had, Florey refused to see them. He thought it would be incorrect to raise people's hopes before he knew how to mass-produce the drug. Fleming alone was seen in the papers and his name has been held high ever since. The Nobel committee got things in a little better proportion when they gave the 1945 Prize jointly to Fleming, Chain, and Florey.

above *This mounted petri dish dating back to 1935, showing penicillin culture, originally belonged to Fleming.*

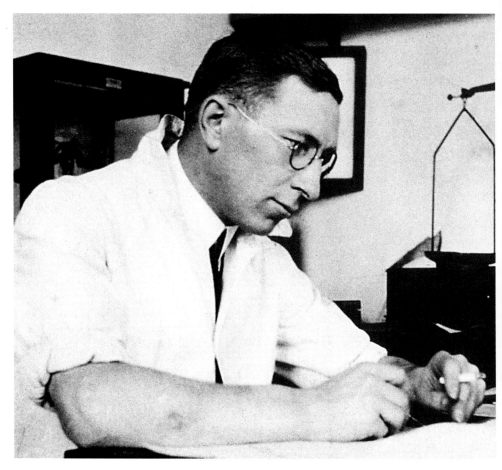

Four thousand years ago the Egyptians noted a pattern of fatal illness. Two thousand years later, Greek philosophers Celsus (c.170) and Araeteus (c.150) started to call it diabetes, but it was not until 1921 that the cause of the illness was discovered and a solution found. Working in Canada, a group of four scientists, Frederick Banting, John Macleod, Charles Best, and Bertram Collip together identified that the pancreas produces a protein that controls the levels of sugar in a mammal's blood. Calling the protein insulin, they discovered how to isolate it and give it to patients. It saves lives by preventing the symptoms of the disease.

Frederick Banting	**Acquaintances**
1891–1941	—Joseph von Mering (1849–1908)
	—Oskar Minkowskin (1858–1931)
John Macleod	—James Bertram Collip (1892–1965)
1876–1935	—Charles Herbert Best (1899–1978)

Frederick Banting & John Macleod

A slow accumulation of evidence

Diabetes has had a long and deadly history. The severe form of the disease, now called Type I diabetes, usually starts to affect people between the ages of 10 and 16, leaving the person with excessively high levels of sugar in their blood. This damages blood vessels, which in turn causes blindness, makes the person lose weight, feel permanently hungry and thirsty, and produce large amounts of urine. Without treatment they soon die.

In the 17th century, English anatomist Thomas Willis recognized that people with this illness had urine that had a sweet taste, and a hundred years after his death others found that the urine was loaded with sugar-like substances. It was the first clue as to the nature of the disease. At the time, most scientists thought that sugar was a dangerous substance to be found in the body, but that was set right by French physiologist Claude Bernard, who in 1857 discovered that the liver turns the starch-like substance glycogen into sugar, which it secretes into the blood. Bernard was well aware of post-mortem results suggesting that the pancreas was damaged in people who had diabetes. In experiments on animals he tried tying off the ducts that led out from the pancreas into the animal's gut, but it had no effect on the animal's sugar levels.

The next link in the chain of events came from two German investigators, Joseph von Mering and Oskar Minkowskin, who in 1889 removed the pancreas from dogs and saw them develop all the symptoms of diabetes. They found that if only part of the pancreas was left behind, or if a slice of its tissue was implanted under the skin, the symptoms did not arise. Clearly, something present in the pancreas that did not pass down into the gut was needed to prevent the onset of diabetes.

In 1869, Paul Langerhans had shown that the pancreas contained clusters of cells that had no duct leading into the gut or other organs, and scientists began to speculate that these "islets of Langerhans" might be secreting chemicals directly into the blood stream.

Removing trypsin reveals the answer

The breakthrough came in the form of a medical demonstrator. Frederick Grant Banting was working as a part-time instructor at the University of Western Ontario's medical school. He was preparing a lecture on the pancreas when he had a bright idea. Reviewing all the evidence, he concluded that people who had tried to find an extract of the pancreas that could remove the symptoms of diabetes had failed because of contamination.

The pancreas secretes an enzyme into the gut that breaks up proteins, and Banting thought that this trypsin might be degrading the vital protein in the laboratory. He approached the professor of physiology at the University of Toronto, John James Rickard Macleod, with a suggestion. If they tied off the ducts from the pancreas, the cells producing trypsin would die. They could then remove the pancreas and extract the mystery protein.

Macleod was initially unimpressed, but with Banting's persuasion he gave in, and in the spring of 1921 Banting started work in his laboratory. The excitement began in the summer while Macleod was on vacation in his native Scotland. Assisted by a medical student, Charles Best, who happened to be looking for a project to work on, Banting removed the pancreas from dogs that had previously had their pancreatic ducts tied-off. He ground up the organs and injected an extract into dogs that had diabetes. The blood sugar level in these animals fell immediately. They repeated the experiment, to be certain that the effects were real. Once again the blood sugar levels dropped dramatically. The next step was to find out what compound in the ground-up organ was responsible. For this they enlisted the support of biochemist J. Bertram Collip of the University of Alberta, who helped them obtain a purer extract of the effective protein, which by now had been called "insulin."

In January 1922 they injected insulin into a patient, 14-year-old Leonard Thompson, whose symptoms of diabetes improved almost miraculously. They had shown clearly that diabetes resulted from a lack of insulin and found a way of preventing the symptoms of the disease. Sadly, they had not found a cure, and indeed that goal has remained elusive.

There was an element of controversy when the Nobel committee gave the 1923 Prize to Macleod and Banting. Banting felt that Macleod had played little or no part in the work and so he shared his part of the prize money with Best. For his part, Macleod also recognized the work done by Collip and gave him half of his winnings. The real winners, however, were the people with diabetes who now have a way of combating the worst effects of their disease.

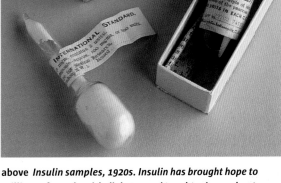

above *Insulin samples, 1920s. Insulin has brought hope to millions of people with diabetes and taught science about hormonal control of the body.*

opposite *Frederick Banting at work in his laboratory.*

October 15, 1951, saw an event that radically altered social structures in the developed world. Working in a small laboratory in Mexico City, Carl Djerassi synthesized a hormone that would eventually be made into a tablet and taken by women to prevent themselves becoming pregnant. The date is in effect the birthday of "the pill." His association with this development has not only influenced his research career, but also caused Djerassi to think about the philosophy of relationships and the way that, in his words, "Sex should be done for pleasure; reproduction for reproduction."

1923–

Acquaintances
— Ludwig Haberlandt (1885–1932)
— Gregory Goodwin Pincus (1903–1967)

Carl Djerassi

The first steps

Tracking the moment when a new scientific idea arrives on the scene is fraught with problems. There is seldom a single moment. More normally, ideas start to germinate, hopes rise and fall, and then eventually something meaningful appears. This was certainly the case with oral contraceptives.

The person who initially suggested that women might be able to take a tablet that could prevent them becoming pregnant was Austrian professor of physiology Ludwig Haberlandt. Working at the University of Innsbruck in 1919, he implanted ovaries from a pregnant rabbit into another non-pregnant rabbit. For the next few months, this rabbit was incapable of becoming pregnant. According to Haberlandt, the animal was temporarily sterilized. Four years later he repeated the work, this time injecting extracts from a placenta into rabbits and mice. Again, the animals could not become pregnant.

As it stood, this concept was far from being a human contraceptive, but it did show that hormones produced in the ovary and placenta during pregnancy prevent the animal producing any more eggs, and consequently make her temporarily infertile. It makes biological sense not to have different-date pregnancies running at the same time, so these hormones were in effect nature's contraceptive. A newspaper headline on January 20, 1927, summed up Haberlandt's hopes: "My aim: fewer but fully desired children!"; and in 1931 he wrote *Die hormonale STERILISIERUNG des weiblichen Organismus*, a book that foretold a day when oral contraceptives would be used for both clinical and eugenic uses.

From hairy yams to oral pills

A few years after Haberlandt's death, chemists in Germany, the United States of America, and Switzerland worked out that the active agent from the ovaries was a steroid that they called progesterone. As it stood, progesterone had two problems — it was expensive to purify, but more importantly the hormone was poorly absorbed and had very little biological activity if you swallowed it in a tablet.

In a slightly bizarre twist of history, Russell Marker, a research professor at Pennsylvania State College, discovered that a particular type of inedible, hairy Mexican yam could yield vast quantities of diosgenin, a chemical that was readily converted into progesterone. This potentially solved the supply problem, but did not get around the fact that the hormone would be destroyed in the stomach if swallowed.

This is where Djerassi comes into the story. While working for the pharmaceutical giant CIBA in 1949, he received a letter inviting him to join Syntex, a tiny company in Mexico. It was the outfit that had made progesterone from yams. He took the appointment and started working on ways of producing steroids from diosgenin. Soon Syntex gained an international reputation for its innovation and successes. It was in this atmosphere that Djerassi recalled papers that he had read a decade earlier while studying for his Ph.D., papers that described ways of altering progesterone molecules, but retaining its function.

On October 15, 1951, he and a young Mexican chemist, Luis Miramontes, synthesized a molecule they called 19-nor-17alpha-ethynyltestosterone, or norethindrone for short. A few days later they packaged some of it up and posted it to Elva G. Shipley, a researcher at Endocrine Laboratories Inc., in Madison, Wisconsin. Shipley soon reported that this molecule mimicked progesterone more powerfully than any other known chemical. Further tests showed that it worked even when taken orally.

The rest is history

At this stage no one was thinking of using the drug as a contraceptive. The initial uses of the molecule were for treating women who had menstrual disorders and who had previously resorted to painfully injected hormones. But then, in 1956, two American researchers, John Rock and Gregory Pincus, saw the potential and conducted the world's first large-scale trial in which synthetic hormones were used as an oral contraceptive. The location of the trial was Puerto Rico and Haiti and 6,000 women took part.

The pill moved rapidly from laboratory to lifestyle and accompanied a change in society's attitude toward sexual relationships. In the following years, millions of women around the world have used oral contraception and Djerassi has developed a sideline in creating literary works. His plays, stories, and poetry encourage people to discuss the way that science is conducted and the effect that it has on our society.

He is well aware that the pill started to alter humanity's role in reproduction, giving people much more control over their involvement in bringing children into existence. He is one of the first to acknowledge that the pill has been just the first step, and more recently reproductive technologies have further increased that control.

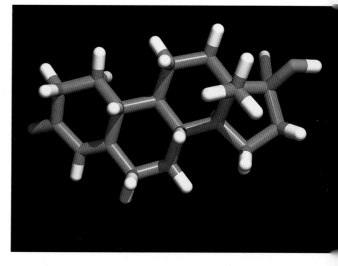

above *Molecule of the contraceptive hormone norethindrone, commonly used in the contraceptive pill. The bonds are color-coded: carbon is pink, hydrogen is white, and oxygen is blue. Norethindrone is taken either by itself, or combined with a synthetic oestrogen.*

The body produces many chemicals that act as messengers that influence the way that particular organs operate. Among these are some that stimulate muscles to contract, and others that cause the muscles to relax. Scientists working in the 1950s were puzzled by adrenaline because this messenger both stimulated and relaxed muscles. James Black believed that these opposing actions resulted from adrenaline acting via two different receptors, so-called alpha- and beta-receptors. Knowing the structure of chemicals that could mimic adrenaline, he looked for new variants that would block the receptors. Beta-blockers were launched into medicine, treating angina, high blood pressure, and many other conditions.

James Black

1924–

Acquaintances

— D'arcy Wentworth Thompson
   (1860–1948)
— George H. Hitchings (1905–1998)
— Gertrude B. Elion (1918–1999)
— William Weipers (died 1990)
— John Robert Vane (1927–)
— Graham John Durant (1934–)
— Charon Robin Ganellin (1934–)
— John Colin Emmett (1939–)

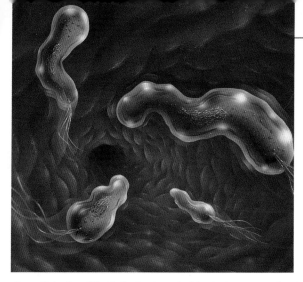

above *Colonies of the* **Helicobacter pylori** *bacteria occur on the mucous membrane of the stomach in people suffering from gastritis. Its presence is thought to increase the risk of tumors.*

Blocking one of two receptors

For centuries philosophers and scientists had no way of knowing how the body worked. Most tended to see it as a spiritual entity, and did not expect to find explicable mechanisms of action driving organs like the heart or the gut. Then, in the 1900s, the mist started to clear and biologists and chemists together found that chemicals and proteins worked inside the body because their shapes allowed them to perform specific tasks.

One of the tasks was to carry messages around the body. Nerves and some organs release chemicals that have specific structures and travel to the part of the body they intend to influence, where their shapes allow them to lock onto receptors. The cell that owns the receptor recognizes this and triggers a specific response. A few of the messenger molecules had more complex actions. Scientists had discovered that adrenaline caused some muscles to contract and others to relax. In itself this did not make sense.

For Black this became a particular issue when he tried to find a way of treating one type of heart complaint: angina pectoris. Clinically, this was triggered by anxiety or exercise and scientists knew that it could also be induced by injecting adrenaline. The problem was that under these conditions the heart was not getting enough oxygen.

There were two ways of solving the oxygen deficit in the heart. One was to increase supply by expanding blood vessels, and the other was to reduce demand by forcing the heart to slow down. At this point, doctors knew that chemicals like nitroglycerine opened up small blood vessels and hoped that if they gave it to their patients, it would increase blood flow within the heart muscle and therefore deliver more oxygen. However, the drugs were not particularly effective.

Consequently, Black looked for ways of slowing the heart down. He knew that adrenaline increased heart rate, so it made sense that blockers of adrenaline would slow it down. So far that hadn't been the case. Blockers that had been tried had relaxed blood vessels, but had no effect on heart rate.

As he looked into the issue, Black became aware of an idea put forward by an American scientist Raymond

Ahlqvist who, in 1948, had proposed that there were two types of adrenaline receptor. He called them alpha- and beta-receptors. Black realized that the effects could be explained if blood vessels contained alpha-receptors, but the heart had beta-receptors, and the blockers that were available were alpha-blockers. He therefore set out to produce some beta-blockers. His strategy was to look at the structure of adrenaline and produce similar chemicals in the hope that one would act selectively on the beta-receptors.

The strategy paid off. In 1958 another group of scientists produced a drug that could block beta-receptors and slow the heart, but this drug had damaging side-effects and so had no clinical value. Then in 1962 Black and his colleagues found that pronethalol slowed the heart and was safe enough to give to patients. In 1964 they had modified it further and come up with propranolol. Clinical trials of propranolol showed that Black's concept was correct, and beta-receptor blocking drugs have subsequently been used throughout the world to treat angina pectoris. They were later found to be helpful in reducing the death rate among people who were recovering from heart failure.

On to histamine

In 1972 Black followed similar lines of reasoning to show that there were two different receptors that respond to the chemical messenger histamine. Producing a blocker of the H2-receptor gave him a drug that reduced acid secretion in the stomach, and formed a major weapon in combating stomach ulcers. Marketed as Tagamet, it has been one of the world's most widely used drugs. Together, these drugs were among the most important contributions to clinical medicine and pharmacology in the 20th century.

Timeline

1924
Born in Uddingston, Scotland
1946
Graduates in medicine at St. Andrew's University, Scotland
1950
Returns to Scotland and starts work in the Department of Physiology in the University of Glasgow
1957
Moves to Singapore to find work
1964
Develops the first clinically useful beta-blocker, propranolol
1965
Moves to Smith, Kleine & French
1972
Characterizes a new group of histamine receptors, H2-receptors, and subsequently develops an H2-blocker, cimetidine, which is used extensively to treat stomach ulcers
1974
Moves to the Wellcome Foundation
1988
Shares the Nobel Prize for Physiology or Medicine with Gertrude B. Elion (1918–1999) and George H. Hitchings (1905–1998)

Antibodies are molecules of protein that are responsible for recognizing foreign molecules that break into the body. These foreigners may vary from bits of dirt to bacteria or viruses. Recognizing them is the first step in a sequence of events designed to remove them. When American Gerald Edelman and Englishman Rodney Porter were working in the 1950s scientists had realized their importance in natural defense, and also that they could be tricked into existence by vaccination. By revealing the structure of these antibodies, Edelman and Porter showed how they work. This has allowed a more rational approach to research in areas of immunology, clinical diagnosis of disease, and therapy.

Gerald Edelman
1929–

Rodney Porter
1917–1985

Acquaintances
— Karl Lansteiner (1868–1943)
— Macfarlane Burnet (1899–1985)
— Arne Wilhelm Kaurin Tiselius (1902–1971)
— Archer John Porter Martin (1910–)
— Niels Kai Jerne (1911–1994)
— Frederick Sanger (1918–)

Gerald Edelman & Rodney Porter

Naturally defensive

The body has its equivalent of an immigration service that operates a very restrictive policy. Nothing foreign should be allowed to stay. In order to operate this rigid regime it needs a mechanism of searching though every nook and cranny, looking for any particles that have just appeared. The way it does this is to use antibodies, and American Gerald Edelman and Englishman Rodney Porter, working independently, came up with data that, when brought together, revealed how these antibodies are made. The knowledge of their structure pointed to an obvious mechanism of action, and the two men subsequently shared the 1972 Nobel Prize for Physiology or Medicine.

The shared prize was fitting as so much of their lives were lived in parallel. They had both served in World War II, Porter in the British armed forces, taking part in the invasion of Algeria in 1942, and later in the invasion of Sicily and Italy, and Edelman was a Captain in the US Army Medical Corps. The war interrupted both of their studies.

Breaking the problem down

Antibodies are formed on the surface of B-lymphocytes, a type of white blood cell, and shed into the body fluids. Each antibody recognizes only one type of foreign particle, so an individual needs to have a vast range of these if they are going to spot all possible invaders.

One of the principal problems that both Edelman and Porter faced was the sheer size of each antibody molecule. It was huge, and there was enormous variation between different molecules in the body.

Porter started by breaking the molecule up into smaller units, hoping to analyze the fragments and then reconstruct the whole molecule. He used papain, an enzyme that cleaves proteins, and found that it chopped the antibody into two types of fragment — Fc readily formed crystals, while Fab had a weakened ability to act as an antibody. Porter realized that for every Fc he had two Fabs and suggested that the molecule consisted of three areas, two of which bound to foreign particles, antigens, and one that didn't.

Edelman took a different approach. He had started from the assumption that antibodies, like many other biologically active molecules, would be made out of collections of chain-like structures linked together by sulfur bonds. He set about finding ways of breaking the sulfur bonds and soon found that each antibody fell apart into two "heavy" chains, and two "light" chains. Unlike Porter's fragments, none of these chains retained any reactivity.

The resolution of this apparently conflicting data came when both scientists realized that antibody molecules were effectively a "Y" shape, but made from four chains. The two "heavy" chains stretched from the top of each arm of the "Y" right down to the bottom. The two "light" chains lay against each of the arms. The chains are held together by sulfur bonds, so Edelman's work had merely separated the chains. Porter's work, on the other hand, had cut the molecule in

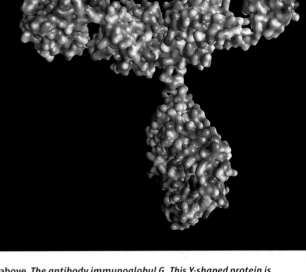

above *The antibody immunoglobul G. This Y-shaped protein is produced by white blood cells as part of an immune response. The "limbs" (upper left and right) bind to foreign antigens such as the surface proteins of invading bacteria.*

opposite *Rodney Porter (left) and Gerald Edelman (right).*

the middle, releasing the two arms of the "Y" and the stem. The arms consisted of part of the heavy and all of the light chain, and this was the section of the molecule that bound to foreign particles.

Wide-ranging applications

This discovery has been revolutionary. It turns out that this basic structure is repeated for all antibodies. It is therefore the key that unlocked our understanding of the science that underlies the body's ability to detect invading organisms. In more recent years, this knowledge has been used to boost a person's immune systems by deliberately stimulating the production of antibodies via vaccination.

The fact that the body increases production of specific antibodies when it encounters particular diseases means that the discovery of antibodies can be used to diagnose illness. If large amounts of a specific antibody are found in an animal's blood, then the animal has encountered that particular disease. Because antibodies bind only to specific targets, scientists are now building them so that they can carry therapeutic drugs to specific target tissues. Edelman and Porter revealed the structure of the molecule, and in so doing have given birth to a massive area of research and industry.

Francis Galton

1822–1911

Acquaintances
— Karl Friedrich Gauss (1777–1855)
— Charles Darwin (1809–1882)
— Karl Pearson (1857–1936)

An explorer and anthropologist, Francis Galton had a passion to measure. He measured distances and temperatures, and worked out the best way to make a cup of tea. His most controversial contribution to human thought occurred when he started measuring human beings and suggesting that a person's physical and mental attributes were inherited from their parents. In doing this he founded the science of eugenics, and set about trying to establish fundamental differences between races. He believed that preventing people with undesirable traits from having children would lead to healthier people living in a safer society.

Nature versus nurture

Born in Birmingham, England, his father was a wealthy banker and his mother was the sister of Charles Darwin's mother. These turned out to be aspects of inheritance that shaped Galton's life, in that his father gave him the ability to live an adventurous life without any need of earning money, while his association with Darwin caused him to think long and hard about principles of inheritance.

From the beginning, Galton was bright. He could read when he was two and a half and by the age of four he was writing and doing arithmetic. At eight he was comfortable reading Latin. He spent the first half of his adult life hunting, shooting, and traveling, and then became captivated by the belief that children inherit all of their characteristics from their parents. He also created the idea that a skilled person should be able to measure some of a person's physical features and make deductions about their innate characteristics and capabilities. He saw little influence coming from the environment in which a person was brought up. For him "nature" was everything, "nurture" was unimportant. He consequently set in motion the "nature versus nurture" debate, and at the same time coined the word "genius."

In his book *Hereditary Genius*, he set out his opinion in characteristically forceful, if long-winded, terms. "I have no patience with the hypothesis occasionally expressed, and often implied, especially in tales written to teach children to be good, that babies are born pretty much alike, and that the sole agencies in creating differences between boy and boy, and man and man, are steady application and moral effort. It is in the most unqualified manner that I object to pretensions of natural equality. The experiences of the nursery, the school, the University, and of professional careers, are a chain of proofs to the contrary."

Measuring mentality

Darwin had proposed that inheritance occurred by "pangenesis," in which "gemmules" in bodily fluids mixed during mating. Galton thought that if this was the case he ought to be able to transfuse blood from one animal to another, and in so doing transfuse the characteristics. Both Galton and Darwin were disappointed when the experiments failed, though Galton then suggested that the study of twins would be another route to discovering the influence of inherited characteristics. Since Galton's day, this has become an intriguing and controversial area of research.

His interest in heredity caused Galton to develop a science that he called "anthropometry." He believed that if he collected data from enough people he would start to see patterns between physical features and mental ability. His interest in mental ability grew from Darwin's theory of the survival of the fittest. He deduced that if two animals with equal physical abilities were competing against each other, then the more intelligent individual would win.

To get measurements from enough people, Galton placed advertisements in newspapers, and 9,000 volunteers paid to attend his Anthropometric Laboratory in London. His problem then was handling the data. But by joining forces with statistician Karl Pearson, he was able to develop some of the earliest statistical tools that look for patterns and correlations within data.

In *Hereditary Genius*, Galton develops a mechanism of ranking individuals, and uses it to claim that there are inherent differences between people from different races. His ideas are summed up in the chapter that is aptly entitled "The Comparative Worth of Different Races." Galton used a letter system to grade various natural abilities. "X" denotes high ability and "x" low ability. The full scale reads: x g f e d c b a A B C D E F G X.

He believed that in any community the average ability should be denoted as between a and A. This would mean that, by definition, there would be the same number above and below. Galton was convinced that the exact value given

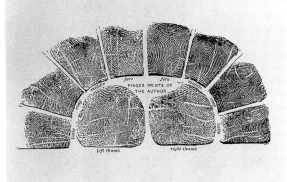

above *Galton was convinced that fingerprints would point out a person's intelligence. Instead they became a valuable forensic tool.*

for that average ability would vary from group to group —
race to race. He believed that a group of "Negroes" would
automatically average two points lower than a group of
Victorian Englishmen. Native Australians scored one point
lower than Negroes and lowland Scotsmen scored one point
higher than the Victorian Englishmen. For Galton, the
superior race were the Athenians, who scored one point
higher than the Scotsmen.

As part of this quest for patterns he developed an
ingenious method of producing photographs that were the
averages of images from 10 different people. He initially
believed that this would show the facial features that
indicated a person would be violent or have criminal
tendencies. This effort was unsuccessful, because the
averaged faces had "no villainy written on them." The
telltale features that he believed were present in individual
faces disappeared in the averaged images.

His work led him to look for a way of increasing the
general level of human intelligence as a means of securing
humanity's future, and with this came the beginnings of
the eugenics movement. In his autobiography *Memories of
My Life*, he tried to draw a distinction between his ideas of
eugenics, and what he felt were misconceptions from
people who thought he wanted to control all human
breeding: "I think that stern compulsion ought to be
exerted to prevent the free propagation of the stock of those
who are seriously afflicted by lunacy, feeble-mindedness,
habitual criminality and pauperism, but that is quite
different from compulsory marriage."

At the beginning of the 20th century, eugenics played an
active role in policy-making in some parts of the United
States and in Nazi Europe, but it is generally felt that the
improvement of populations via restricted breeding is
inappropriate. Some people believe that, as we enter the
21st century, a new form of eugenics is emerging as a
consequence of the selection of embryos in reproductive
technologies and genetic testing during pregnancy.

Timeline

1822
Born on February 16 into a wealthy family, he is Charles
Darwin's cousin
1840
Enters Trinity College, Cambridge
1845
Receives a large inheritance when his father dies and
consequently never needs to earn money
1850
Decides to explore Africa
1853
In the same year as he gets married, he is elected as a fellow of
the Royal Geographical Society
1859
Charles Darwin publishes *The Origin of Species*
1860
Galton makes significant advances in the sciences of mapping
and weather forecasting
1865
Performs his first research into eugenics and publishes his ideas
in an article for *Macmillan's Magazine*
1869
Publishes *Hereditary Genius*
1908
Publishes his autobiography, *Memories of My Life*
1909
Galton is knighted
1911
Dies on January 17

below *Francis Galton's Second Anthropometric Laboratory at
the International Health Exhibition in London (1891–1895).
Anthropometry is the scientific study of the measurements
and proportions of the human body.*

In the 1800s there was a growing realization that if a person's brain was physically deformed or damaged, they could suffer from a mental illness. Working in a Parisian mental asylum, Jean-Martin Charcot turned the care of people with mental disorders into a science. He carefully observed the behavior of individual patients and then correlated these with detailed studies of their brains, nerves, and muscles when they had died. His observations led him to distinguish between diseases that were caused by a physical impairment, and those that were more based on disturbed emotions.

Jean-Martin Charcot

1825–1893

Acquaintances
— Guillaume Benjamin Amand Duchenne (1806–1875)
— John Hughlings Jackson (1835–1911)
— Charles-Joseph Bouchard (1837–1915)
— George Miller Beard (1839–1883)
— Pierre Marrie (1853–1940)
— Howard H. Tooth (1856–1926)
— Sigmund Freud (1856–1939)

Physical damage in a gunpowder plot

In 1862, at the age of 37, Jean-Martin Charcot started work as the senior physician at Salpêtrière hospital in Paris. This famous hospital on the Left Bank of the River Seine was named as a reminder of the fact that it had once been an arsenal and gunpowder store for Louis XIII, but it now served as a hostel for more than 5,000 patients. Charcot had been born in Paris on November 29, 1825, and as a boy spent many hours drawing and painting—hobbies that taught him to make careful observations and records.

Having become a junior doctor in the hospital when he initially qualified in 1853, he had risen steadily through the ranks. The mental hospital gave Charcot a remarkable level of access to observe his patients while they were alive and study their bodies after they had died.

Contemporary opinion was that nervous and mental disorders were all due to physical damage to parts of the patient's brain. Charcot collected data for this clinical-anatomical method of research. Working with Charles-Joseph Bouchard, he established that nerve cells within the spinal cord degenerated in some forms of neurological disease, expanding the area of study from simply looking for damage within the brain. He became convinced that epileptic fits were triggered from highly localized areas in an individual's brain, building on ideas initially presented by English neurologist John Hughlings Jackson.

Charcot did not limit his work to the hospital. He deliberately employed a housemaid because she had disseminated sclerosis. Having her work in his house meant that he could study her closely, even if it did mean that he lost many cups and plates. In a somewhat macabre manner, he followed his observations through to the post-mortem examination of her body.

At the same time as observing illness, Charcot was able to establish the function of zones within healthy brains. For example, he was the first person to work out which parts of the brain organized muscle control.

Hysteria doesn't fit the bill

The hospital had many young women who fainted and had convulsions, contortions, and transient lapses of consciousness. The women were housed together so that they could be looked after, but also so that they could be studied. Charcot was convinced that this group of women had a unique disorder, that he called hystero-epilepsy. One of his students, Joseph Babinski, was less convinced. He argued that Charcot had in effect invented this so-called hystero-epilepsy. He pointed out that the women had arrived with vague symptoms of distress and demoralization, but that as they had moved under Charcot's care, they had developed the symptoms that were displayed by the other women on the ward. Babinski believed that the

below *Charcot's demonstrations of hysteria were regular events in Paris attracting philosophers, medics, and aristocracy who wanted to see the spectacle.*

women's real problem was that they were depressed and vulnerable to suggestion.

Having listened to his student, Charcot tried a new approach to treating these women. First, they were moved to normal wards around the hospital so that they were separated from other similarly behaving people. The staff on these wards were told to ignore, rather than investigate, any of these symptoms. Without the role models and the attention, the effect was dramatic: the symptoms disappeared. Staff concentrated on giving the patients physiotherapy and discussing conflicts they had at home.

The episode was an important step for Charcot and for neurology. Apart from anything, it showed that not all neurological disorders were the result of physically damaged nervous systems.

Duchenne and his magic electric box

As well as his clinical work, Charcot ran regular teaching sessions and lectures. His presentations had more than a touch of theatre, being delivered on a floodlit stage in the amphitheatre of the Salpêtrière. He often involved patients, bringing them onto the platform to demonstrate some observation or clinical feature. Among his audience were notable names, including Sigmund Freud and Guillaume Benjamin Amand Duchenne. Freud claims that it was these lectures that stimulated his interest in finding a psychological rather than a physical origin of neurosis. Freud was also impressed by Charcot's use of hypnotism, a technique that Charcot believed could only be used on people with hysteria.

Duchenne attended from time to time, as much as a friend and colleague as a student. He had been born in the French seaside town of Boulogne and had studied at Douai and Paris. Having turned his back on the seafaring tradition of his family, he entered medicine. After suffering from depression when his wife died, Duchenne virtually abandoned his work, but then became interested in the effect of electricity on people's skin.

He built a portable machine that contained an induction coil capable of generating short pulses of electricity, and in the 1840s started wandering around Parisian hospitals. In a technique he called "faradism" he stimulated the nerves and muscles of patients, partly trying to cure them and partly trying to map the muscles of the body and note their functions.

By applying electricity, Duchenne found that moving the arm or a finger required not just one muscle to contract, but a series of muscles to act together. At the same time he was able to distinguish between paralysis caused by brain damage and that caused by faulty muscles. If the problem was in the brain then his faradism would make the muscles contract, but if the muscle or its nerves were injured then the electrical stimulation had little or no effect.

In 1850, Duchenne published his first paper describing the way that the muscles in a person's face allow him or her to create facial expressions. His work had been conducted both on patients and on bodies just after the people had died, but their muscles still responded to electricity. By

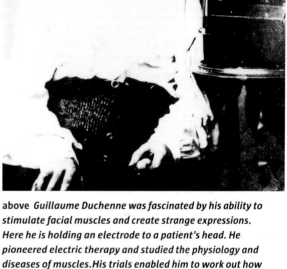

above *Guillaume Duchenne was fascinated by his ability to stimulate facial muscles and create strange expressions. Here he is holding an electrode to a patient's head. He pioneered electric therapy and studied the physiology and diseases of muscles. His trials enabled him to work out how muscles in the face operated.*

1862 he had mapped 100 facial muscles and noted that half-hearted smiles involved only the muscles around the mouth, but genuine smiles, showing "the sweet emotions of the soul," also involved muscles around the person's eyes.

While collecting his data he encountered a nine-year-old boy who was losing his ability to walk, due to muscle-wasting. In 1868, Duchenne published an extensive account of this disease, drawing on observations of 12 other cases. As the first symptoms of the disease were an apparent over-development of the boy's calf muscles Duchenne called it pseudohypertrophis dystrophy, but it eventually became more commonly known as Duchenne muscular dystrophy, or D.M.D.

Between them, Charcot and Duchenne had demonstrated the need of studying disease in living subjects, and by carefully observing healthy and diseased nervous systems they had created the discipline of neurology.

Ivan Pavlov

1849–1936

Acquaintances
— John Watson (1878–1958)
— Dimitri Ivanovich Pisarev (1840–1868)
— Ivan Mikhailovich Sechenov
  (1829–1905)
— Sergei Petrovich Botkin (1832–1889)

Science often investigates living organisms by taking them to pieces and working out how the individual parts work. A biochemist may look at the way an enzyme functions and an immunologist may study antibodies, glands, and white blood cells. But for Ivan Pavlov, the only way to make sense of an animal was to look at the whole organism. By looking at animals as complex sets of neurally-interconnected organs he revealed many of the physical processes that underpin subjective experiences, such as emotion, learning, and personality. His work set the foundations for the controversial study of behaviorism.

From theology to physiology

Ever since Galileo Galilei and Isaac Newton, the physical sciences had become areas of experimental research and mathematics. But midway through the 19th century, research into the workings of living animals was still dominated by philosophical concepts such as the will, desire, and reason. Ivan Pavlov was one of a group of scientists who changed that and started to perform careful measurements and scientific experiments on complex living organisms.

Having been born the son of a priest, Pavlov was initially destined to become a priest himself. His early education was in religious schools, but ones that fostered an interest in natural science. Inspired by the progressive ideas of the 19th century Russian literary critic Dimitri Pisarev and the father of Russian physiology Ivan Mikhailovich Sechenov, Pavlov abandoned his religious career and in 1870 enrolled in the physics and mathematics faculty at the University of St. Petersburg, to take a course in natural science.

He excelled in scholarship and research and during his career developed three overlapping strands of investigation: he studied the heart and circulation of blood, the digestive system, and complex nervous functions in the brain.

Saliva—a window on the brain

In Pavlov's day there was much less concern about the treatment of animals than there is in the 21st century, and no one had heard of an animal rights movement. The consequence is that some of his work may seem outrageous to a modern observer, but it would have raised little, if any, anxiety at the time he did it.

In order to understand how a body worked, Pavlov decided that you needed not only to look at an entire animal, but also to observe it for many days, weeks, or months. If this observation was going to be accurate and measurable, it would mean surgically implanting measuring equipment into the animal, or at times surgically altering part of the animal's anatomy. This was new: previously scientists had performed short experiments on animals that disrupted the normal physiological processes.

The concept of anesthesia was only just arriving on the scene, but Pavlov realized that it altered the way that the animals behaved, so rather than using it he trained his animals to lie still during surgery. In fact, Pavlov showed an unusually high degree of concern for his animals, knowing that the results from his experiments would have meaning only if they were healthy and comfortable.

Central to Pavlov's quest was the desire to find out how the brain influenced the way that animals interact with their environment. To study this he started by looking at saliva, and what he termed "psychic" salivary secretion. He had noticed that dogs sometimes salivate for no apparent reason, and decided to investigate the process. Some theorists suggested that the dogs were salivating because they wanted to have food, but Pavlov was not convinced. Pavlov believed that the dog's salivary glands were stimulated by an external signal.

To test this, he started by playing a metronome near a dog. Nothing happened. The regular tick, tick, tick was a "neutral" stimulus. Placing ground-up meat next to the animal made it salivate. Pavlov called this an "unconditional" stimulus. He then played the metronome just before giving the dog some meat. Soon the dog would start salivating when the metronome played and before the meat arrived.

In 1903, he explained to the 14th International Medical Congress in Madrid that the metronome was no longer "neutral," but had become a "conditioned" stimulus—Pavlov said that the dog had become "sensitized" to the metronome. He had demonstrated that training an animal could cause the formation of new connections within the brain and showed the way that reflex responses to the environment drive many of an animal's actions.

Three principles

The starting point of Pavlov's work had been Sechenov's theoretical attempt to discover how reflex mechanisms operate. Pavlov had provided experimental proof of the process. He then went on to refine the ideas and developed three principles for the theory of reflexes. The first was the principle of determinism, which establishes that many of an animal's actions are determined by a conditioned reflex. The second was the principle of analysis and synthesis, which suggested that the animal must be able to collect and integrate information from many different sources; and the third was the principle of structure, which established that there must be some physical means in the body for making all this occur.

Developing these principles led him and his colleagues to build up a scientific theory of the laws that govern how an organism functions as a whole. Pavlov's team became convinced that conditioned reflexes originated in the cerebral cortex, the intricately folded outer layer of the cerebrum, which in humans forms some 40 percent of the weight of the brain. He called this the "prime distributor and organizer of all activity of the organism."

above *The laboratory of Ivan Pavlov, 1904, with a group of laboratory staff and dogs in the grounds. Pavlov knew that reliable results would only come from healthy animals.*

While many commentators were not comfortable with the idea that animals, and especially human beings, were driven purely by complex interactions of reflexes, others acclaimed Pavlov's work. They claimed it was wonderful science and used it to form the background to behaviorism, the school of psychology that seeks to explain animal and human behavior entirely in terms of observable and measurable responses to environmental stimuli. Ironically, though Pavlov decried the communist movement, his ideas fitted in well with the developing communist ideology. Consequently, after the October Revolution, a special government decree signed by Lenin on January 24, 1921, stated that "the outstanding scientific services of Academician I.P. Pavlov … are of enormous significance to the working class of the whole world." The Communist Party gave him almost unlimited scope for research, causing the Soviet Union to become a prominent center for the study of physiology.

Pavlov had shown the importance of nervous reflex responses, but more importantly he established a new mindset and the means to perform physiological experiments that revealed the way that organisms operate as a whole.

Timeline

1849
Born in Ryazan, a small village in central Russia on September 14

1879
Receives his Ph.D. from the University of St. Petersburg, after studying chemistry and physiology

1881
Marries a fellow student, Serafima Karchevskaya, and the couple subsequently have four sons, one of whom dies as a child, and a daughter

1895
Becomes professor of pharmacology at St. Petersburg Institute of Experimental Medicine and then switches to become the professor of physiology

1903
Publishes his famous Conditioning Reflex

1904
Receives the Nobel Prize for Physiology or Medicine for his research on the digestive system

1913
His ideas play a large role in the behaviorist theory of psychology introduced by John Watson (1878–1958)

1936
Dies on February 27

One and a half centuries after his death there is still a great rift between those who hold Sigmund Freud in admiration and those who pour disdain on his work. Freud explored the way that people's minds operate. He developed a set of theories based around the idea that all humans have unconscious processes that drive their thinking and actions. Through psychoanalysis he claimed to discover that the two most potent unconscious influences were sexual and aggressive drives. He caused scandal by claiming that erotic desire starts in infancy, and in so doing coined the Oedipus complex.

Sigmund Freud

1856–1939

Acquaintances
— Jean-Martin Charcot (1825–1893)
— Josef Breuer (1842–1925)
— Alfred Adler (1870–1937)
— Carl Gustav Jung (1875–1961)
— Albert Einstein (1879–1955)

It's all in the mind

Scientific study and research were fundamental parts of Sigmund Freud's medical education in Vienna, and throughout his life Freud was passionate that he should be seen as a scientist, especially when challenged by friends and colleagues. Shortly after returning from spending a year with French neurologist Jean-Martin Charcot in the mental hospital in Paris, Freud established his own private practice, aiming to treat people with psychological disorders.

Initially he tried using Charcot's techniques of hypnotism to treat people with hysteria, but found that its beneficial effects did not last. Instead, he turned to a method suggested by fellow physician Josef Breuer, who had found that if patients talked freely about the earliest occurrences of their symptoms, the hysteria gradually disappeared.

Together, Freud and Breuer developed a theory that phobias, hysterical outbursts, pains, and some forms of paranoia were caused by traumatic events that had occurred earlier in the patient's life. The key thing was that, in normal existence, the patient would not remember these events. They were held only in the person's subconscious. Freud would ask his patients to lie down on a couch and say whatever came into their heads, and he found that they spoke more freely if he was out of sight. The rationale was that a person would relax, forget that the analyst was present, and consequently speak freely about his or her experiences and feelings.

By using this psychoanalysis, Freud hoped to reveal the subconscious neuroses and allow the person to confront them and subsequently gain control. This thinking was revolutionary, but fitted in with other changes in scientific thought. Charles Darwin had recently driven out the notion that humans were totally unique from the rest of the animal world. This inadvertently introduced the possibility that, like the behaviors of other animals, human behavior and thought could be subject to scientific investigation. For Freud, this meant that human behavior must be determined not by a person's soul, but by an individual's previous experiences, and the only way to make sense of this was to presume that each person must have a subconscious part of his or her mind. The mind is then likened to an iceberg where the visible part, the conscious, sits on top of a much larger subconscious region.

Freud believed that the mind was driven by two basic categories of instincts. One was Eros, the group of self-preserving instincts, which includes sexual activity. The

above *Sigmund Freud with Carl Jung, Ernest Jones, Sandor Ferenczi and two unidentified others, 1909.*

other was Thanatos, a group of instincts that urge us towards aggression, self-destruction, and cruelty. Of the two, he frequently asserted that Eros was the group with greatest influence.

A structured mind

In 1923, Freud developed a more complex view of the mind and, like Plato, he divided the mind into three structural elements. While Plato divided mind into wisdom, courage, and self-control, Freud chose id, ego, and super-ego. The id is the part of the mind that operates the drives that need to be satisfied, such as sexual drives. The super-ego is that part that contains conscience, and is highly influenced by social norms and parental upbringing. The ego is the conscious area of mental activity that has the task of reconciling the id and the super-ego.

According to Freud, a person will have a stable mental health if the id and the super-ego are held in balance. Problems occur when, for example, the id needs some source of satisfaction that the super-ego maintains is banned by society. If the ego can't find a way of resolving the conflict then the person may suffer from some neurosis.

To defend itself from these neuroses, Freud maintained that the mind adopts a number of different strategies. It may repress issues by pushing conflicts back into the subconscious, it may sublimate them by, for example, channeling sexual drives into socially acceptable achievements like sport. In addition the mind could fixate and fail to progress beyond one of the developmental stages, or regress and return to a more infantile behavior.

Dreams and slips of the tongue started to take on great significance for Freud, because he believed they were moments when repressed thoughts momentarily broke loose from the subconscious. Using his tripartite view of the mind, the task of psychoanalysis was to locate the issues that the super-ego had repressed and hidden within the id and bring them into consciousness, thus allowing the ego to confront and discharge them.

The Oedipus complex

If Freud had shocked people by suggesting that they were repressing their sexuality, he scandalized them by insisting that this repression started as an infant. While working with Breuer he became convinced that any traumatic events experienced by children would powerfully influence them as adults. He then developed the idea that childhood sexual experiences critically determine adult personality, and at this point Freud's focus on sex as the primary drive caused a split between himself and Breuer.

Working with Carl Jung and Alfred Adler, Freud started to expand this theme, but when Freud insisted that sexual conflict started in infancy his collaborators again deserted him. Freud believed that children followed a set pattern of development in the way that they satisfy the sexual demands of their id. Initially the children have an "oral" phase and satisfy the demand by sucking. They then move to an "anal" stage, where the demand is met by pleasure in

the act of defecation. As they pass through the "phallic" stage they develop acute interest in their sexual organs. From as early as the age of five the child moves on to acquire a deep sexual attraction for the parent of the opposite sex. Freud termed this last stage the "Oedipus complex," after the mythical story of Oedipus who unwittingly kills his father and marries his mother.

Many of Freud's ideas developed from a period of extensive self-analysis undertaken in the hope of understanding some of his own behavior and mood swings, and his thoughts were consequently shaped by the complexity of his own family relations. His mother was a second wife. He grew up alongside a stepbrother's son who was the same age as himself and was the source of much friendship and jealousy, and, in addition, he had a younger brother die in infancy. All this left Freud with aggressive feelings towards his father.

While the details of Freud's life are complex and his ideas are controversial, he does however stand in history as the person who established the idea that we can look into the mind to see why people behave as they do.

Timeline

1856
Born on May 6, in Freiburg, Moravia (now Pribor in the Czech Republic)

1860
His family moves to Vienna

1881
He receives his medical degree from the University of Vienna

1885
Spends a year in Paris studying with Jean-Martin Charcot at the Salpêtrière Hospital

1886
Begins private neurology in Vienna and marries Martha Bernays. The couple subsequently have six children

1896
Coins the term "psychoanalysis"

1900
Publishes *The Interpretation of Dreams*

1908
Holds the First International Congress of the Vienna Psychoanalytic Society

1909
Gives lectures in the United States of America which boost his international reputation

1910
Establishes the International Psychoanalytic Association

1923
Publishes *The Ego and the Id*

1938
When Austria is annexed by Germany, he leaves Vienna for London

1939
Dies in London of cancer on September 23

In the autumn of 1904 the French government had a problem. It had recently passed a new law requiring that all French children should be given an education. Faced with the task of teaching all children, the schools soon found that some did not appear to be as capable of learning as were others. Concerned about the issue of helping these low-achieving children, the government asked Alfred Binet to devise a test that could pick out slow learners. The result was the first test that roughly estimated a child's intelligence, and despite Binet's opposition to Intelligence Quotients it became the forerunner of the IQ test.

Alfred Binet

1857–1911

Acquaintances
— John Stuart Mill (1803–1873)
— Jean-Martin Charcot (1825–1893)
— Wilhelm Max Wundt (1832–1920)
— Théodore Simon (1883–1961)

Building on Galton

In the late 19th century, Francis Galton made pioneering studies of individual differences in intelligence. He believed that people with good senses of hearing, seeing, smell, touch, and taste would be highly intelligent. Moreover, because he was convinced that these abilities were inherited, he believed that individual differences in intelligence were also inherited. In his book *Hereditary Genius*, Galton argued that eminent fathers tended to have eminent sons.

In 1888, Galton had established his "anthropometric laboratory" in order to measure physical features of people and assess their intelligence. Working in the United States, psychologist James McKeen Cattell referred to these as "mental tests," but by 1901 this Galtonian approach was being abandoned because no association had been found between physical features and intellectual performance.

Against this backdrop, French psychologist Alfred Binet began his work on intelligence scales. Binet had studied hypnosis, thinking, and individual differences, but his first attempts to measure intelligence were based on measuring the size of children's heads. Having found no discernible patterns between the head sizes of talented and poorly performing students, he moved on to argue that individual differences in intelligence could only be detected by measuring complex processes such as memory, imagination, attention, comprehension, and suggestibility. After a somewhat checkered career in which he found it difficult to hold the same job for very long, Binet was appointed in 1904 by the French minister of public instruction to develop tests that could measure intelligence in children.

Urbanization and industrialization were introducing new challenges, among which was the need to provide everyone with formal education. The French government and school administrators wanted to reorganize their schools, and wanted to identify children who would be slow learners so that they could be given specialist help.

The Binet-Simon scale

Consequently, Binet developed a scale that could differentiate slow learners from those who were able to keep pace with normal levels of instruction. Working with Théodore Simon, a young physician who had experience studying retarded children, he constructed a series of specific tests. Binet initially observed his two daughters, but, working with Simon, he extended the scope of study to cover a large group of school children. Their goal was to find a spectrum of tests that could clearly separate children into a normal and a slow group. They ended up with a set of 30 tests.

In 1905, Binet and Simon published their rationale and their tests, and argued that any discussion of the cause of a mental retardation — nature or nurture — was irrelevant. The tests were designed to avoid anything resembling schoolwork and included activities like unwrapping food, remembering shopping lists, and placing different weights in correct order. "It is the intelligence alone that we seek to measure, by disregarding in so far as possible the degree of instruction which the child possesses ... We give him nothing to read, nothing to write and submit him to no test in which he might succeed by means of rote learning," they said.

This psychological method measured comprehension, judgment, reasoning, and invention, and gave an indication of the child's general intelligence at that moment. Binet believed that the tests reflected the nature of intelligence, as they revealed the child's practical ability to adapt to new circumstances. He was also convinced that a person's intelligence could change over time, and he consequently developed a technique of "mental orthopedics," that he claimed could increase the mental levels of children with both normal and retarded abilities.

Age matters

One useful feature of the tests was that an individual child's score could be compared with the average score for that age. In 1908, Binet and Simon introduced age into their scale. They said that if, for example, 75 percent or more of six-year-old children could pass a particular test, that test was placed at the six-year level. In addition, a six-year-old child who performed as well as the average eight-year-old would have a "mental level" of eight.

When Binet's writings were translated into English, the term "mental age" was used, implying an ordered developmental progression that Binet had not intended. He argued against German psychologist Wilhelm Max Wundt, who had suggested that you could calculate an Intelligence Quotient, an IQ, by comparing a child's actual achievement with the norm for that age. Binet thought that the reduction of intelligence to a single value was far too simplistic, but he died in 1911, too soon to pursue that argument.

Timeline

1857
Born in Nice, France, on July 11
1883
Studies hypnosis with pioneering neurologist Jean-Martin Charcot at the Salpêtrière Hospital in Paris
1884
Marries Laure Balbiani and the couple subsequently have two daughters
1890
Breaks off connections with the Salpêtrière and concentrates on studying his two daughters
1891
Binet takes an unpaid job at the Sorbonne
1898
Binet is joined by student Théodore Simon
1905
Publishes *The Binet-Simon Test of Intelligence*
1911
Shortly after revising his tests, Binet dies in Paris

Are human beings essentially spiritual in nature or machines that follow patterns of behavior that are largely determined by complex interactions of conditioned responses? For John Watson, the answer was simple. In establishing the concept of behaviorism, he said that all animals, including humans, were complex machines that respond to situations according to the way that their brains are "wired" and the experience that has conditioned their minds. He believed that this understanding of behavior could lead to ways of treating people who had mental disorders. After being forced out of academia he took his ideas of behaviorism into the advertising industry.

John Watson

1858–1935

Acquaintances
—Aldof Meyer (1866–1950)
—James Rowland Angell (1869–1949)
—Ivan Pavlov (1849–1936)
—Sigmund Freud (1856–1939)

An unlikely start

At the start of the 20th century Sigmund Freud was creating a stir in psychological circles by introducing psychoanalysis. Freud was suggesting that by going through a process of introspective questioning, a person could reveal deeply-hidden damaging experiences, and that once identified they could be dealt with. The concept was highly acclaimed by some and derided by others. John Watson became one of the people who derided it.

Having been born into a family where his mother was deeply religious and banned dancing, drinking, and smoking, while his father conducted a series of affairs with other women, the young Watson grew up to be a difficult and, at times, violent teenager. At school he mocked teachers, got into fights, and assaulted black children. Out of school he had a few scrapes with the local police.

The opportunity to go to Furman University in Southern Carolina, however, offered him a fresh start and an absorbing intellectual challenge. There he was introduced to psychology and went on to study it for a Ph.D. at the University of Chicago. He disliked using human subjects in experiments, becoming convinced that he could find out all he needed by looking at animals. After all, he reasoned, humans are only more complex versions of other animals.

His work was good and five years after getting his Ph.D., Johns Hopkins University appointed him professor of experimental and comparative psychology.

Behaviorism is born

By the time that he arrived at the Johns Hopkins University, Watson had developed the foundations for a new idea. He was inspired by Ivan Pavlov's work and goaded by that of Freud, saying that Pavlov's conditional responses could make better sense of human behavior than Freud's introspection, which Watson saw as bordering on mystic.

In 1913, Watson presented his manifesto "Psychology as the Behaviorist Views It." In this he reviews the previous "obvious" failings of introspection and offers an alternative definition of psychology as the "science of behavior." Watson felt that in studying what people do, psychology was in a position to make predictions about a person's behavior and reveal ways that that behavior could be controlled.

He hoped to turn psychology into a true science, though his detractors point out that even though he moved from Freud's introspection, he made no attempt to explain the physical processes in the brain that underlie his findings. Neither did he set up testable theories in a way that is common to all other areas of science.

World War I interrupted his studies, but when he returned to academic life he started looking at ways of conditioning and controlling people's emotions. Part of this study included the Little Albert experiment. In this he presented a boy called Albert with a series of animals, making loud noises each time they arrived. The child soon displayed fear even when the animals arrived without the noise. This and other evidence convinced Watson that all children had three basic emotional reactions: fear, rage, and love.

Selling on fear, rage, and love

The controversy over an affair with one of his students forced Watson to leave academia. After a brief pause he moved into advertising. He found that executives had learned from the propaganda successes of World War I and had become scientific "consumption engineers." Working at the J. Walter Thompson Agency, Watson had the opportunity of putting some of his ideas about controlling human behavior into practice. He believed that if an advertiser worked on the three basic emotions of fear, rage, and love, it would grab people's attention and force them to respond.

To this end, he started to move advertising away from its preoccupation with simply presenting information about the product, to becoming a medium that appeals to non-rational emotions. His success seems to have been modest at best, though this is probably largely because he was ahead of his time.

Timeline

1878
Born in Greenville, South Carolina
1891
His father leaves home after a string of extra-marital affairs
1903
Receives his Ph.D. in psychology from the University of Chicago
1905
Marries Mary Ickes, and the couple subsequently have two children, Mary and John
1913
Presents "Psychology as the Behaviorist Views It" in his presidential address to the American Psychological Association
1920
Leaves academia after having an affair with Rosalie Rayner, one of his research students. He subsequently marries Rosalie and they have two sons, James and William
1924
Tries to implement behaviorism while working as vice-president at J. Walter Thompson, one of the United States' largest advertising agencies
1958
Dies on his farm in Connecticut

Carl Jung

1875–1961

Acquaintances
— Sigmund Freud (1856–1939)
— Eugen Bleuler (1857–1939)

The 18th-century philosopher Immanuel Kant introduced the idea that the human mind actively created our perception of the world, rather than simply being a passive recipient of experiences. Building on this, Carl Jung created a theory of a three-part mind, consisting of the ego, the personal unconscious, and a collective unconscious. He claimed that people could be divided into introverts and extroverts, and believed that the collective unconscious was built of archetypes that we inherit, and is satisfied by activities such as religion and special relationships. In recent years his work has formed the basis of assessments like the Myers-Briggs personality test.

A double personality

The Swiss psychiatrist Carl Jung ascribed many of his beliefs to his family and initial upbringing. His father was a pastor in the Reformed Church, but had lost his faith. His mother's parents were spiritualists, and she claimed to have a dual personality. In the daytime she was "hearty with animal warmth," but at night she became much more aware of the supernatural. In his autobiographical book *Memories, Dreams, Reflections*, Jung wrote that he too shared this double life, having what he called his rational number one personality that was interested in academic and worldly success, and his unconscious number two personality that he felt was irrational and in tune with the paranormal.

Seeking to make sense of his two personalities, Jung went to medical school and specialized in psychiatry. After graduating he became the assistant to Eugen Bleuler who worked at the clinic in Zürich, a renowned authority in schizophrenia, and got to know psychoanalyst Emma Rauschenbach. The two got on so well intellectually and personally that they got married.

A two-phase relationship with Freud

Out of his strange mixture of personal contacts, Jung developed a theory of word association. He asked patients to say what words came into their minds when another key word is given, believing that this revealed complexes within the person's unconscious. Jung's word association impressed Freud, and the two started corresponding by letter in April 1906. The correspondence grew and in March 1907 Jung traveled to Vienna to meet Freud. On their first encounter they talked constantly for 13 hours, and Jung commented later that "nobody within my own area of experience was capable to measure with Freud."

For his part, Freud was equally impressed with Jung, and saw him as his natural successor in leading the growing psychoanalytic movement. As a consequence of Freud's patronage, Jung became the first president of the International Psychoanalytic Association in March 1909, but the friendship didn't last much longer.

Freud warned Jung to defend psychoanalysis from what he saw as the tidal mud of occultism. Jung, however, had a personal interest in the paranormal. He became unhappy with Freud's obsession with the sexual basis of all behavior: Freud's assertion that this sexuality commenced in infancy was the final straw. The two parted company in a very public row that stretched from 1912 to 1914.

Types and archetypes

In a similar way to Freud's psychoanalysis, Jung based his analytical psychology on the idea that a person's mind has unconscious and conscious elements. He then went on to divide people between two groups depending on his assessment of their personality types—extrovert and introvert. He worked on the idea during an intense four- or five-year period that followed his break with Freud. This was a time of profound mental anguish for Jung that wasn't helped by the fact that he had an affair with a former patient and a consequent row with his wife.

Trying to resolve his inner conflict by confronting his unconscious, Jung studied his dreams intensely, and entered into dialogue with his "inner figures" or archetypes. According to Jung, archetypes are instinctive parts of every human being. These archetypes are expressed in the form of art, religion, mythology, astrology, and folklore, and Jung at times referred to them as dominants, and mythological or primordial images.

Jung claimed that there are as many archetypes as there are typical situations in life. He saw them in some ways like empty picture frames, which needed filling with content. Content enters the frame as a person has an experience that relates to that specific archetype. As a consequence, the exact content may vary from person to person, culture to culture, but the same set of archetypal frames are present in every human.

For Jung, these archetypes formed a collective unconscious, a psychic inheritance that is in effect a reservoir of our experiences as a species. It is a "knowledge" that each person is born with, but being part of our unconscious we can never be directly aware of it. Jung believed that this collective unconscious influences all of our experiences and behavior, especially those relating to our emotions. He felt that previously inexplicable experiences such as déjà vu and love at first sight were glimpses of this feature at work.

A key example is the mother archetype. All our ancestors must have had mothers. Jung would say that our evolutionary development has given us a built-in ability to recognize a mothering relationship. The mother archetype is abstract, and needs a physical representation in the form of a particular person, usually a person's mother. He saw this archetype symbolized by the primordial earth mother of mythology, or by Eve and Mary in Christianity, as well as less personal symbols such as the Church, the nation, or even an ocean.

Jung thought that a person who had a poor relationship with his mother would satisfy the need to fill this archetype by seeking comfort in the Church or identifying with a "motherland" — he even suggested that some might seek this fulfillment by spending their lives at sea.

Jung also talked of "mana," an archetype that is represented in many cultures by some form of phallic symbol. Unlike Freud, he did not see this as a representation of sex, but of power and fertility. In addition, the shadow was another archetype that was a hangover from our pre-human animal past, and was concerned with survival and reproduction.

Set against these collective archetypes, Jung thought that everyone also had a personal unconscious. This is like a reference library, in that it holds both memories that are easily brought to mind and those that have been suppressed for some reason, but can be revealed by particular cues or triggers. Like Freud, Jung considered that ego formed the third part of the psyche, where ego was basically the conscious part of the mind.

From Greece to modernity

Following these years of experimenting with the unconscious, Jung expanded his interest in world religions through various anthropological expeditions. He studied gnosticism, mythology, and medieval alchemy.

Jung's legacy has been a mass of literature that has inspired people involved in creativity, spirituality, and psychic phenomena. On the other hand, scientists have been more cautious about his work. To start with, he seeks to describe the mind from the basis of how it ought to work, rather than a rational approach of looking to see what is physically present. In this way he is reuniting his concept of science with the early Greek philosophers. Also, while science tends to use a process of reductionism to break things down to the smallest unit and then study that unit in the hope of one day putting the whole system back together, Jung's concept was the opposite. He wanted to look at highest levels of organization, such as the mind, and then make theories about its smaller components.

Jung did, however, give a valuable appreciation of the difference between child and adult development, saying that children emphasize differences between objects, while adults seek meaning by finding how different things fit together. His theories have also given way to assessments like the Myers-Briggs test, that play a critical part in many businesses, as they enable individuals to become aware of themselves, and allow people to work effectively in teams and groups.

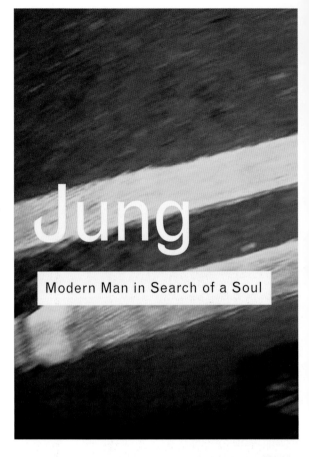

Jung

Modern Man in Search of a Soul

Timeline

1875
Born in Kesswil, Switzerland, on July 26
1900
Graduates with a medical degree from the University of Basel
1902
Receives his Ph.D. while working with Eugen Bleuler at the Burghoelzli Mental Hospital in Zurich
1903
Marries psychoanalyst Emma Rauschenbach and starts looking at word association
1906
Starts corresponding with Sigmund Freud
1907
Visits Freud and is deeply impressed by his intellect
1909
Opens his private practice in Kuessnacht, which he runs until his death. In the same year he takes a joint lecture tour of America with Freud
1912
Publicly splits from Freud during a lecture tour in America
1919
Starts using the term "archetype"
1961
Dies at Kuessnacht

top right *Like many of Jung's writings, this book looks at his ideas about analyzing dreams, the primitive unconscious, and the relationship between psychology and religion.*

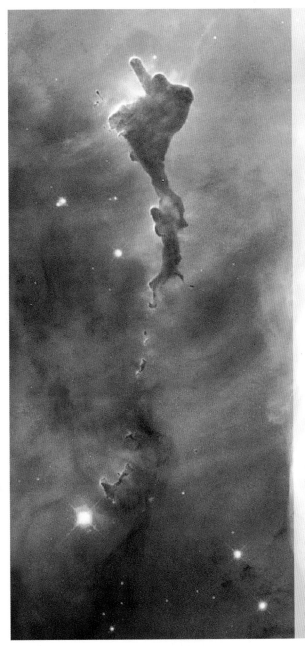

The ancient Greek philosophers would be thrilled to see what has become of their opening salvos in the war on ignorance. Plato talked about the philosopher who broke free the shackles that confined him to a world of dancing shadows, walked out of the cave, and saw the true light. The process of science has been, and continues to be, one where individual thinkers have escaped from conventional views of life, the universe, and everything, and presented novel explanations of reality.

The problem with Plato's cave is that you can never be certain that you are looking at the true source of light — the ultimate understanding. It is always possible that current thinking is but a shadow of the real thing. New experiments and observations may reveal a more robust understanding, banishing the shadows with increased illumination. This book has shown plenty of instances where one person's theory has become the next person's source of humor.

It is all too easy to slip into the habit of talking about scientific facts, or laws of nature, as if we have revealed some ultimate truths. In reality, these facts and laws are simply the best way that we have of representing our data. It would be arrogant and dangerous to assume that contemporary concepts are rock solid and will not be prone to change.

The majority of people highlighted in this book are no longer alive. That is not to imply that contemporary scientists are in some way inferior, just that without the test of time it is difficult to see which revelations will really shape our futures. You could assume that big-budget projects like particle physics and genetics will be the main drivers of change, but history shows it is not always areas that receive most funding that produce novel breakthroughs. After all, Einstein's initial period of critical thought occurred while he had no funding — when he was working as a patent clerk.

One assumption is safe: human beings have inquiring minds and nothing will stop their expeditions. The consequences will often be unexpected and will frequently challenge our understandings of our universe, our world, and ourselves. We may choose to ignore the science, but we will never avoid the consequences of the revelation.

Conclusion

Index

Picture credits